5

GLOBAL AGRICULTURE: DEVELOPMENTS, ISSUES, AND RESEARCH

COMPARISONS IN HOUSEHOLD FOOD SECURITY

UNITED STATES AND INDIA

Global Agriculture: Developments, Issues, and Research

Additional books in this series can be found on Nova's website under the Series tab.

Additional E-books in this series can be found on Nova's website under the E-book tab.

Nutrition and Diet Research Progress

Additional books in this series can be found on Nova's website under the Series tab.

Additional E-books in this series can be found on Nova's website under the E-book tab.

GLOBAL AGRICULTURE: DEVELOPMENTS, ISSUES, AND RESEARCH

COMPARISONS IN HOUSEHOLD FOOD SECURITY

UNITED STATES AND INDIA

SAMUEL OWUSU
EDITOR

New York

LIBRARY OF CONGRESS CATALOGING-IN-PUBLICATION DATA

ISBN: 978-1-62417-530-5

Published by Nova Science Publishers, Inc. † New York

CONTENTS

PREFACE

This book provides an overview of research in the study of household food security in the United States and India. An estimated 85.1 percent of American households were food secure throughout the entire year in 2011, meaning that they had access at all times to enough food for an active, healthy life for all household members. The remaining households (14.9 percent) were food insecure, meaning that the food intake of one or more household members was reduced and their eating patterns were disrupted at times during the year because the household lacked money and other resources for food. Comparatively, this book provides a quantitative assessment of food security using a large household-level expenditure survey conducted by the Government of India during 2004/2005. The analysis tests the impact of several key assumptions required to estimate actual calories consumed from the expenditure data.

Chapter 1 – An estimated 85.1 percent of American households were food secure throughout the entire year in 2011, meaning that they had access at all times to enough food for an active, healthy life for all household members. The remaining households (14.9 percent) were food insecure at least some time during the year, including 5.7 percent with very low food security—meaning that the food intake of one or more household members was reduced and their eating patterns were disrupted at times during the year because the household lacked money and other resources for food. The prevalence rate of very low food security increased from 5.4 percent in 2010, returning to the level observed in 2008 and 2009. The change in food insecurity overall (from 14.5 percent in 2010) was not statistically significant. The typical food-secure household spent 24 percent more on food than the typical food-insecure household of the same size and household composition. Fifty-seven percent of all food-insecure households participated in one or more of the three largest Federal food and nutrition assistance programs during the month prior to the 2011 survey.

Chapter 2 – This study provides a quantitative assessment of food security using a large household-level expenditure survey conducted by the Government of India during 2004/05. The analysis tests the impact of several key assumptions required to estimate actual calories consumed from the expenditure data. The authors found significant differences in the estimates of calories consumed and the number of food-insecure people under alternative plausible assumptions for computing the calorie content of nonprocessed foods, processed foods, and meals eaten outside the household. The measurement errors were largest in accounting for calories consumed by the highest and lowest income households. Overall, the difference between the highest and lowest estimate of the number of people consuming an average of less than 2,100 calories per day was equivalent to about 17 percent of India's population, or 173 million people in 2004/05. Given the significant measurement error in estimating calories consumed, it is important to consider not only consumption surveys, but also aggregate food availability studies and survey data on anthropometric measures that accompany undernourishment—such as growth stunting—in assessing food insecurity.

In: Comparisons in Household Food Security ISBN: 978-1-62417-530-5
Editor: Samuel Owusu

Chapter 1

HOUSEHOLD FOOD SECURITY IN THE UNITED STATES IN 2011*

Alisha Coleman-Jensen, Mark Nord, Margaret Andrews and Steven Carlson

ABSTRACT

An estimated 85.1 percent of American households were food secure throughout the entire year in 2011, meaning that they had access at all times to enough food for an active, healthy life for all household members. The remaining households (14.9 percent) were food insecure at least some time during the year, including 5.7 percent with very low food security—meaning that the food intake of one or more household members was reduced and their eating patterns were disrupted at times during the year because the household lacked money and other resources for food. The prevalence rate of very low food security increased from 5.4 percent in 2010, returning to the level observed in 2008 and 2009. The change in food insecurity overall (from 14.5 percent in 2010) was not statistically significant. The typical food-secure household spent 24 percent more on food than the typical food-insecure household of the

* This is an edited, reformatted and augmented version of the United States Department of Agriculture Publication, ERR-141, dated September, 2012.

same size and household composition. Fifty-seven percent of all food-insecure households participated in one or more of the three largest Federal food and nutrition assistance programs during the month prior to the 2011 survey.

Keywords: Food security, food insecurity, food spending, food pantry, soup kitchen, emergency kitchen, material well-being, SNAP, Food Stamp Program, National School Lunch Program, WIC

SUMMARY

What Is the Issue?

Most U.S. households have consistent, dependable access to enough food for active, healthy living—they are food secure. But a minority of American households experience food insecurity at times during the year, meaning that their access to adequate food is limited by a lack of money and other resources. Food and nutrition assistance programs of the U.S. Department of Agriculture (USDA) increase food security by providing low-income households access to food, a healthful diet, and nutrition education.

USDA also monitors the extent and severity of food insecurity in U.S. households through an annual, nationally representative survey sponsored by USDA's Economic Research Service. Reliable monitoring of food security contributes to the effective operation of the Federal programs as well as private food assistance programs and other government initiatives aimed at reducing food insecurity. This report presents statistics from the survey covering households' food security, food expenditures, and use of food and nutrition assistance programs in 2011.

What Did the Study Find?

The percentage of U.S. households that were food insecure remained essentially unchanged from 2010 to 2011, while the percentage with food insecurity in the severe range—described as very low food security—increased.

- In 2011, 85.1 percent of U.S. households were *food secure* throughout the year. The remaining 14.9 percent (17.9 million households) were *food insecure*. Food-insecure households (those with low and very low food security) had difficulty at some time during the year providing enough food for all their members due to a lack of resources. The change from the 2010 estimate (14.5 percent) was not statistically significant, meaning that the difference may be due to sampling variation.
- In 2011, 5.7 percent of U.S. households (6.8 million households and one-third of all food-insecure households) had *very low food security*. In these households, the food intake of some household members was reduced and normal eating patterns were disrupted at times during the year due to limited resources. The prevalence of very low food security returned to the level observed in 2008 and 2009, a statistically significant increase from the 5.4-percent level of 2010. Increases in the prevalence of very low food security were greatest for women living alone, Black households, and households with annual incomes below 185 percent of the poverty line.
- Children were food insecure at times during the year in 10.0 percent of households with children (3.9 million households), essentially unchanged from 9.8 percent in 2010. These households were unable at times during the year to provide adequate, nutritious food for their children.
- While children are usually shielded from the disrupted eating patterns and reduced food intake that characterize very low food security, both children and adults experienced instances of very low food security in 1.0 percent of households with children (374,000 households) in 2011, unchanged from 2010.
- For households with incomes near or below the Federal poverty line, households with children headed by single women or single men, and Black and Hispanic households, rates of food insecurity were substantially higher than the national average. Food insecurity was more common in large cities and rural areas than in suburban areas and other outlying areas around large cities.
- Typically, households classified as having very low food security experienced the condition in 7 months of the year, for a few days in each of those months.

- The typical food-secure household spent 24 percent more for food than the typical food-insecure household of the same size and composition, including food purchased with Supplemental Nutrition Assistance Program (SNAP) benefits (formerly called food stamps).
- Fifty-seven percent of food-insecure households in the survey reported that in the previous month, they had participated in one or more of the three largest Federal food and nutrition assistance programs.

How Was the Study Conducted?

Data for the ERS food security reports come from an annual survey conducted by the U.S. Census Bureau as a supplement to the monthly Current Population Survey. USDA's Economic Research Service sponsors the annual survey and compiles and analyzes the responses. The 2011 food security survey covered 43,770 households comprising a representative sample of the U.S. civilian population of 119 million households. The food security survey asked one adult respondent in each household a series of questions about experiences and behaviors that indicate food insecurity, such as being unable to afford balanced meals, cutting the size of meals because of too little money for food, or being hungry because of too little money for food. The food security status of the household was assigned based on the number of food-insecure conditions reported.

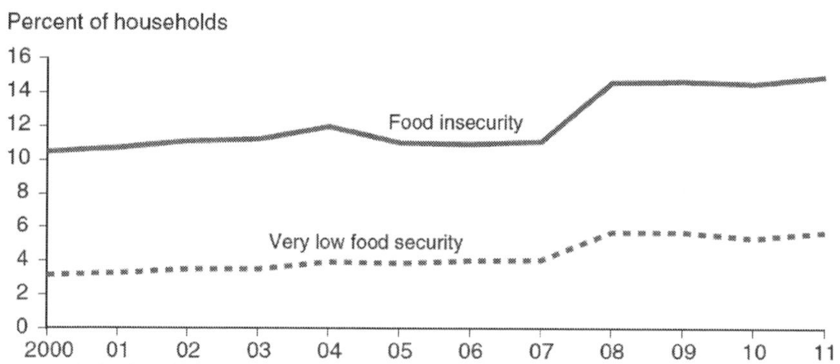

Source: Calculated by USDA, Economic Research Service based on Current Population Survey Food Security Supplement data.

The prevalence of food insecurity changed little since 2008-09.

INTRODUCTION

Since 1995, the U.S. Department of Agriculture has collected information annually on food access and adequacy, food spending, and sources of food assistance for the U.S. population. The information is collected in an annual food security survey, conducted by the U.S. Census Bureau as a supplement to the nationally representative Current Population Survey.[1] A major impetus for this data collection is to provide information about the prevalence and severity of food insecurity in U.S. households. Previous reports in the series are available at: http://www.ers.usda.gov/topics/food-nutrition-assistance/food-security-in-the-us/readings.aspx/.

This report updates the national statistics on food security, household food spending, and the use of Federal food and nutrition assistance by food-insecure households, using data collected in the December 2011 food security survey—the 17th annual survey in the Nation's food security monitoring system. Additional statistics, including the prevalence of food insecurity during the 30 days prior to the food security survey, the frequency of occurrence of food-insecure conditions, and use of food pantries and emergency kitchens, are available online at http://www.ers.usda.gov/publications/ap-administrative-publication/ap-058.aspx/.

HOUSEHOLD FOOD SECURITY

Food security—access by all people at all times to enough food for an active, healthy life—is one of several conditions necessary for a population to be healthy and well nourished. This section provides information on food security and food insecurity in U.S. households over the course of the year ending in December 2011.

Methods

The statistics presented in this report are based on data collected in a supplement to the Current Population Survey (CPS) conducted in December 2011. The CPS currently includes about 53,000 households and is representative, at State and national levels, of the civilian, noninstitutionalized population of the United States. In December 2011, 43,770 households completed the food security supplement; the remainder was unable or unwilling to do so. Survey sample weights were calculated by the U.S. Census Bureau to indicate how many households were represented by each household that responded to the survey.[2] All statistics in this report were calculated by applying the food security supplement weights to responses by the surveyed households so the statistics are nationally representative.

Unless otherwise noted, statistical differences described in the text are significant at the 90-percent confidence level.[3]

The household food security statistics presented in this report are based on a measure of food security calculated from responses to a series of questions about conditions and behaviors that characterize households when they are having difficulty meeting basic food needs.[4] Each question asks whether the condition or behavior occurred at any time during the previous 12 months and specifies a lack of money and other resources to obtain food as the reason. Voluntary fasting or dieting to lose weight is thereby excluded from the measure. The series includes three questions about food conditions of the household as a whole and seven about food conditions of adults in the household and, if there are children present in the household, an additional eight questions about their food conditions (see box, "Questions Used To Assess the Food Security of Households in the CPS Food Security Survey," page 3). Responses to the 18 food security questions are reported in tables S-6 to S-8 of the Statistical Supplement, http://www.ers.usda.gov/publications/ap-administrative-publication/ap-058.aspx/. The food security status of each interviewed household is determined by the number of food-insecure conditions and behaviors the household reports. Households are classified as food secure if they report no food-insecure conditions or if they report only one or two food-insecure conditions. (Food-insecure conditions are indicated by responses of "often" or "sometimes" to questions 1-3 and 11-13; "almost every month" or "some months but not every month" to questions 5, 10, and 17; and "yes" to the other questions.) They are classified as food insecure if they report three or more food-insecure conditions.[5]

Food-insecure households are further classified as having either *low food security* or *very low food security*.[6] The very low food security category identifies households in which food intake of one or more members was reduced and eating patterns disrupted because of insufficient money and other resources for food. Households without children are classified as having *very low food security* if they report six or more food-insecure conditions. Households with children are classified as having *very low food security* if they report eight or more food-insecure conditions, including conditions among both adults and children (see box, "What Is 'Very Low Food Security'?" on page 5). They are further classified as having *very low food security among children* if they report five or more food-insecure conditions among the children (that is, if they respond affirmatively to five or more of questions 11-18).

QUESTIONS USED TO ASSESS THE FOOD SECURITY OF HOUSEHOLDS IN THE CPS FOOD SECURITY SURVEY

1. "We worried whether our food would run out before we got money to buy more." Was that often, sometimes, or never true for you in the last 12 months?
2. "The food that we bought just didn't last and we didn't have money to get more." Was that often, sometimes, or never true for you in the last 12 months?
3. "We couldn't afford to eat balanced meals." Was that often, sometimes, or never true for you in the last 12 months?
4. In the last 12 months, did you or other adults in the household ever cut the size of your meals or skip meals because there wasn't enough money for food? (Yes/No)
5. (If yes to question 4) How often did this happen—almost every month, some months but not every month, or in only 1 or 2 months?
6. In the last 12 months, did you ever eat less than you felt you should because there wasn't enough money for food? (Yes/No)
7. In the last 12 months, were you ever hungry, but didn't eat, because there wasn't enough money for food? (Yes/No)
8. In the last 12 months, did you lose weight because there wasn't enough money for food? (Yes/No)
9. In the last 12 months did you or other adults in your household ever not eat for a whole day because there wasn't enough money for food? (Yes/No)
10. (If yes to question 9) How often did this happen—almost every month, some months but not every month, or in only 1 or 2 months?

(***Questions 11-18 were asked only if the household included children age 0-17***)

11. "We relied on only a few kinds of low-cost food to feed our children because we were running out of money to buy food." Was that often, sometimes, or never true for you in the last 12 months?
12. "We couldn't feed our children a balanced meal, because we couldn't afford that." Was that often, sometimes, or never true for you in the last 12 months?
13. "The children were not eating enough because we just couldn't afford enough food." Was that often, sometimes, or never true for you in the last 12 months?
14. In the last 12 months, did you ever cut the size of any of the children's meals because there wasn't enough money for food? (Yes/No)
15. In the last 12 months, were the children ever hungry but you just couldn't afford more food? (Yes/No)
16. In the last 12 months, did any of the children ever skip a meal because there wasn't enough money for food? (Yes/No)
17. (If yes to question 16) How often did this happen—almost every month, some months but not every month, or in only 1 or 2 months?
18. In the last 12 months did any of the children ever not eat for a whole day because there wasn't enough money for food? (Yes/No)

Low and very low food security differ in the extent and character of the adjustments the household makes to its eating patterns and food intake. Households classified as having *low food security* have reported multiple indications of food access problems, but typically have reported few, if any, indications of reduced food intake. Those classified as having *very low food security* have reported multiple indications of reduced food intake and disrupted eating patterns due to inadequate resources for food. In most, but not all households with *very low food security*, the survey respondent reported that he or she was hungry at some time during the year but did not eat because there was not enough money for food.

Prevalence of Food Insecurity—National Conditions and Trends

An estimated 85.1 percent of U.S. households were food secure throughout the entire year in 2011 (figure 1, table 1A). In concept, "food secure" means that all household members had access at all times to enough food for an active, healthy life (Anderson, 1990).[7] The remaining 14.9 percent (17.9 million households) were food insecure at some time during the year. That is, they were, at times, unable to acquire adequate food for one or more household members because they had insufficient money and other resources for food. About two-thirds of food-insecure households avoided substantial reductions or disruptions in food intake, in many cases by relying on a few basic foods and reducing variety in their diets. But 5.7 percent (6.8 million households) had very low food security—that is, they were food insecure to the extent that eating patterns of one or more household members were disrupted and their food intake reduced, at least some time during the year, because they could not afford enough food.

Among U.S. households with children under age 18, 79.4 percent were food secure in 2011. In 10.6 percent, only adults were food insecure (figure 2). Both children and adults were food insecure in 10.0 percent of households with children (3.9 million households), and in about 1.0 percent (374,000 households), one or more child was also subject to reduced food intake and disrupted eating patterns at some time during the year (table 1B) In some households with very low food security among children, only older children may have experienced the more severe effects of food insecurity while younger children were protected from those effects (Nord, 2009a).

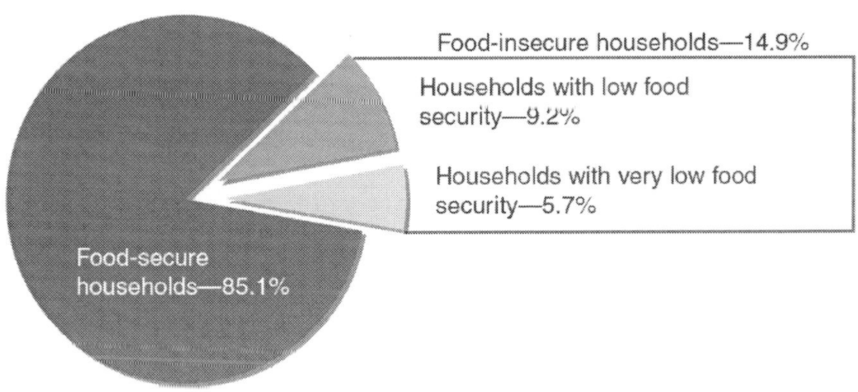

Source: Calculated by USDA, Economic Research Service using data from the December 2011 Current Population Survey Food Security Supplement.

Figure 1. U.S. households by food security status, 2011.

Table 1A. Households and individuals by food security status of household, 1998-2011

Category and year	Total[1]	Food secure		Food insecure					
				All		With low food security		With very low food security	
	1,000	*1,000*	*Percent*	*1,000*	*Percent*	*1,000*	*Percent*	*1,000*	*Percent*
Households:									
2011	119,484	101,631	85.1	17,853	14.9	11,014	9.2	6,839	5.7
2010	118,756	101,527	85.5	17,229	14.5	10,872	9.1	6,357	5.4
2009	118,174	100,820	85.3	17,354	14.7	10,601	9.0	6,753	5.7
2008	117,565	100,416	85.4	17,149	14.6	10,426	8.9	6,723	5.7
2007	117,100	104,089	88.9	13,011	11.1	8,262	7.0	4,749	4.1
2006	115,609	102,961	89.1	12,648	10.9	8,031	6.9	4,617	4.0
2005	114,437	101,851	89.0	12,586	11.0	8,158	7.1	4,428	3.9
2004	112,967	99,473	88.1	13,494	11.9	9,045	8.0	4,449	3.9
2003	112,214	99,631	88.8	12,583	11.2	8,663	7.7	3,920	3.5
2002	108,601	96,543	88.9	12,058	11.1	8,259	7.6	3,799	3.5
2001	107,824	96,303	89.3	11,521	10.7	8,010	7.4	3,511	3.3
2000	106,043	94,942	89.5	11,101	10.5	7,786	7.3	3,315	3.1
1999	104,684	94,154	89.9	10,529	10.1	7,420	7.1	3,109	3.0
1998	103,309	91,121	88.2	12,188	11.8	8,353	8.1	3,835	3.7
All individuals (by food security status of household):[2]									
2011	305,893	255,773	83.6	50,120	16.4	33,232	10.9	16,888	5.5

Category and year	Total[1]	Food secure		Food insecure					
				All		With low food security		With very low food security	
	1,000	*1,000*	*Percent*	*1,000*	*Percent*	*1,000*	*Percent*	*1,000*	*Percent*
2010	304,034	255,202	83.9	48,832	16.1	32,777	10.8	16,055	5.3
2009	301,750	251,588	83.4	50,162	16.6	32,499	10.8	17,663	5.9
2008	299,567	250,459	83.6	49,108	16.4	31,824	10.6	17,284	5.8
2007	297,042	260,813	87.8	36,229	12.2	24,287	8.2	11,942	4.0
2006	294,010	258,495	87.9	35,515	12.1	24,395	8.3	11,120	3.8
2005	291,501	256,373	87.9	35,128	12.1	24,349	8.4	10,779	3.7
2004	288,603	250,407	86.8	38,196	13.2	27,535	9.5	10,661	3.7
2003	286,410	250,155	87.3	36,255	12.7	26,622	9.3	9,633	3.4
2002	279,035	244,133	87.5	34,902	12.5	25,517	9.1	9,385	3.4
2001	276,661	243,019	87.8	33,642	12.2	24,628	8.9	9,014	3.3
2000	273,685	240,454	87.9	33,231	12.1	24,708	9.0	8,523	3.1
1999	270,318	239,304	88.5	31,015	11.5	23,237	8.6	7,779	2.9
1998	268,366	232,219	86.5	36,147	13.5	26,290	9.8	9,857	3.7
Adults (by food security status of household):[2]									
2011	231,385	197,923	85.5	33,462	14.5	21,371	9.2	12,091	5.2
2010	229,129	196,505	85.8	32,624	14.2	21,357	9.3	11,267	4.9
2009	227,543	194,579	85.5	32,964	14.5	20,741	9.1	12,223	5.4
2008	225,461	193,026	85.6	32,435	14.4	20,320	9.0	12,115	5.4
2007	223,467	199,672	89.4	23,795	10.6	15,602	7.0	8,193	3.7

Table 1A. (Continued)

Category and year	Total[1]	Food secure		Food insecure					
				All		With low food security		With very low food security	
	1,000	*1,000*	*Percent*	*1,000*	*Percent*	*1,000*	*Percent*	*1,000*	*Percent*
2006	220,423	197,536	89.6	22,887	10.4	15,193	6.9	7,694	3.5
2005	217,897	195,172	89.6	22,725	10.4	15,146	7.0	7,579	3.5
2004	215,564	191,236	88.7	24,328	11.3	16,946	7.9	7,382	3.4
2003	213,441	190,451	89.2	22,990	10.8	16,358	7.7	6,632	3.1
2002	206,493	184,718	89.5	21,775	10.5	15,486	7.5	6,289	3.0
2001	204,340	183,398	89.8	20,942	10.2	14,879	7.3	6,063	3.0
2000	201,922	181,586	89.9	20,336	10.1	14,763	7.3	5,573	2.8
1999	198,900	179,960	90.5	18,941	9.5	13,869	7.0	5,072	2.5
1998	197,084	174,964	88.8	22,120	11.2	15,632	7.9	6,488	3.3

[1] Totals exclude households for which food security status is unknown because they did not give a valid response to any of the questions in the food security scale. In 2011, these exclusions represented 405,000 households (0.3 percent of all households).

[2] The food security survey measures food security status at the household level. Not all individuals residing in food-insecure households were directly affected by the households' food insecurity. Similarly, not all individuals in households classified as having very low food security were subject to the reductions in food intake and disruptions in eating patterns that characterize this condition. Young children, in particular, are often protected from effects of the households' food insecurity.

Sources: Calculated by USDA, Economic Research Service using Current Population Survey Food Security Supplement data.

Table 1B. Households with children by food security status, and children by food security status of household, 1998-2011

Category and year	Total[1]	Food-secure households		Food-insecure households[2]		Households with food-insecure children[3]		Households with very low food security among children	
	1,000	*1,000*	*Percent*	*1,000*	*Percent*	*1,000*	*Percent*	*1,000*	*Percent*
Households with children:									
2011	38,803	30,814	79.4	7,989	20.6	3,862	10.0	374	1.0
2010	39,419	31,447	79.8	7,972	20.2	3,861	9.8	386	1.0
2009	39,525	31,114	78.7	8,411	21.3	4,208	10.6	469	1.2
2008	39,699	31,364	79.0	8,335	21.0	4,361	11.0	506	1.3
2007	39,390	33,160	84.2	6,230	15.8	3,273	8.3	323	.8
2006	39,436	33,279	84.4	6,157	15.6	3,312	8.4	221	.6
2005	39,501	33,404	84.4	6,197	15.6	3,244	8.2	270	.7
2004	39,990	32,967	82.4	7,023	17.6	3,808	9.5	274	.7
2003	40,286	33,575	83.3	6,711	16.7	3,606	9.0	207	.5
2002	38,647	32,267	83.5	6,380	16.5	3,456	8.9	265	.7
2001	38,330	32,141	83.9	6,189	16.1	3,225	8.4	211	.6
2000	38,113	31,942	83.8	6,171	16.2	3,282	8.6	255	.7
1999	37,884	32,290	85.2	5,594	14.8	3,089	8.2	219	.6
1998	38,036	31,335	82.4	6,701	17.6	3,627	9.5	331	.9
Children (by food security status of household):[4]									
2011	74,508	57,850	77.6	16,658	22.4	8,565	11.5	845	1.1
2010	74,905	58,697	78.4	16,208	21.6	8,458	11.3	976	1.3

Table 1B. (Continued)

Category and year	Total[1]	Food-secure households		Food-insecure households[2]		Households with food-insecure children[3]		Households with very low food security among children	
	1,000	*1,000*	*Percent*	*1,000*	*Percent*	*1,000*	*Percent*	*1,000*	*Percent*
2009	74,207	57,010	76.8	17,197	23.2	8,957	12.1	988	1.3
2008	74,106	57,433	77.5	16,673	22.5	9,098	12.3	1,077	1.5
2007	73,575	61,140	83.1	12,435	16.9	6,766	9.2	691	.9
2006	73,587	60,959	82.8	12,628	17.2	7,065	9.6	430	.6
2005	73,604	61,201	83.1	12,403	16.9	6,718	9.1	606	.8
2004	73,039	59,171	81.0	13,868	19.0	7,823	10.7	545	.7
2003	72,969	59,704	81.8	13,265	18.2	7,388	10.1	420	.6
2002	72,542	59,415	81.9	13,127	18.1	7,397	10.2	567	.8
2001	72,321	59,620	82.4	12,701	17.6	6,866	9.5	467	.6
2000	71,763	58,867	82.0	12,896	18.0	7,018	9.8	562	.8
1999	71,418	59,344	83.1	12,074	16.9	6,996	9.8	511	.7
1998	71,282	57,255	80.3	14,027	19.7	7,840	11.0	716	1.0

[1] Totals exclude households for which food security status is unknown because they did not give a valid response to any of the questions in the food security scale. In 2011, these exclusions represented 154,000 households (0.4 percent of all households with children).

[2] Food-insecure households are those with low or very low food security among adults or children or both.

[3] Households with food-insecure children are those with low or very low food security among children.

[4] The food security survey measures food security status at the household level. Not all children residing in food-insecure households were directly affected by the households' food insecurity. Similarly, not all children in households classified as having very low food security among children were subject to the reductions in food intake and disruptions in eating patterns that characterize this condition. Young children, in particular, are often protected from effects of the households' food insecurity.

Sources: Calculated by USDA, Economic Research Service using Current Population Survey Food Security Supplement data.

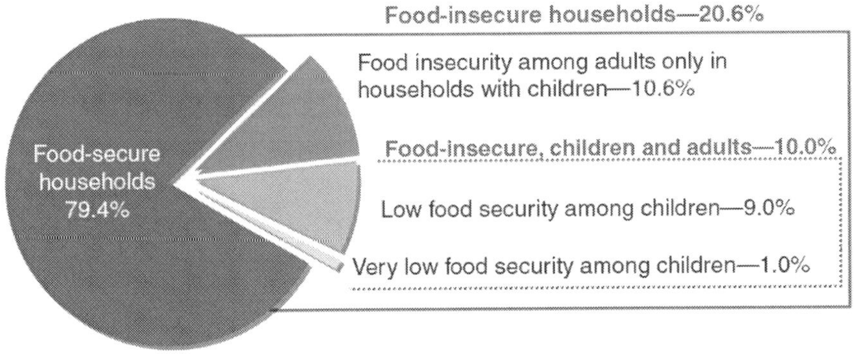

Source: Calculated by USDA, Economic Research Service using data from the December 2011 Current Population Survey Food Security Supplement.

Figure 2. U.S. households with children by food security status of adults and children, 2011.

When interpreting food security statistics in this report, it is important to keep in mind that households were classified as having low or very low food security if they experienced the condition at any time during the previous 12 months. The prevalence of these conditions on any given day is far below the corresponding annual prevalence. For example, the prevalence of very low food security on an average day during the 30-day period prior to the December 2011 survey is estimated to have been between 0.8 and 1.1 percent of households (1.0 million to 1.3 million households; see box, "When Food Insecurity Occurs in U.S. Households, It Is Usually Recurrent But Not Chronic," on page 9). Children, along with adults, experienced very low food security in an estimated 48,000 to 68,000 households (0.12 to 0.17 percent of all U.S. households with children) on an average day during the same period.

The prevalence of food insecurity was essentially unchanged from 2010 to 2011. That is, the change from 2010 was within the range that could have resulted from sampling variation. Over the previous decade, food insecurity had increased from 10.5 percent in 2000 to nearly 12 percent in 2004, declined to 11 percent in 2005-07, then increased to its current level in 2008 (14.6 percent) and remained essentially unchanged at that level in 2009 and 2010 (figure 3).[8]

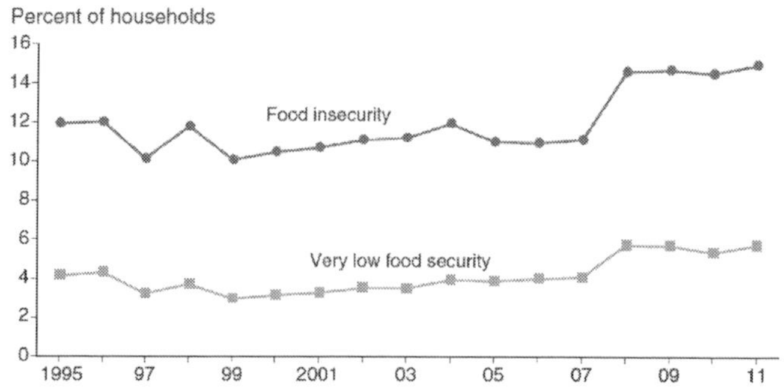

[1] Prevalence rates for 1996 and 1997 were adjusted for the estimated effects of differences in data collection screening protocols used in those years.

Source: Calculated by USDA, Economic Research Service based on Current Population Survey Food Security Supplement data.

Figure 3. Trends in the prevalence of food insecurity and very low food security in U.S. households, 1995-2011[1].

What Is "Very Low Food Security"?

Very low food security can be characterized in terms of the conditions that households in this category reported in the food security survey (households without children classified as having very low food security reported six or more food-insecure conditions and households with children reported eight or more food-insecure conditions, including conditions among both adults and children). Thus, the defining characteristic of "very low food security" is that, at times during the year, the food intake of household members was reduced and their normal eating patterns were disrupted because the household lacked money and other resources for food. In the 2011 survey, households classified as having very low food security (representing an estimated 6.8 million households nationwide) reported the following specific conditions:

- 99 percent reported having worried that their food would run out before they got money to buy more.
- 97 percent reported that the food they bought just did not last and they did not have money to get more.
- 95 percent reported that they could not afford to eat balanced meals.
- 97 percent reported that an adult had cut the size of meals or skipped meals because there was not enough money for food.

- 91 percent reported that this had occurred in 3 or more months.
- 95 percent reported that they had eaten less than they felt they should because there was not enough money for food.
- 65 percent reported that they had been hungry but did not eat because they could not afford enough food.
- 48 percent reported having lost weight because they did not have enough money for food.
- 27 percent reported that an adult did not eat for a whole day because there was not enough money for food.
- 21 percent reported that this had occurred in 3 or more months.
- All households without children reported at least six of these conditions, and 66 percent reported seven or more. (Conditions in households with children were similar, but the reported food-insecure conditions of both adults and children were taken into account.)

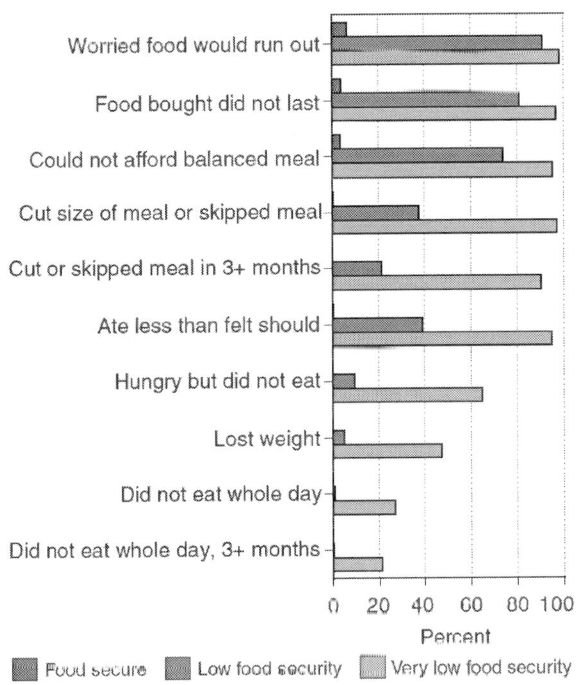

Source: Calculated by USDA, Economic Research Service using data from the December 2011 Current Population Survey Food Security Supplement.

Percentage of households reporting each indicator of food insecurity, by food security status, 2011.

The prevalence of very low food security increased from 5.4 percent in 2010 to 5.7 percent in 2011, returning to the level observed in 2008 and 2009. Prior to that, the prevalence of very low food security had increased from about 3 percent in 2000 to 4 percent in 2004, and remained essentially unchanged through 2007.

Prevalence of Food Insecurity by Selected Household Characteristics

The prevalence of food insecurity varied considerably among households with different demographic and economic characteristics (table 2). Food insecurity was strongly associated with income. For example, 41.1 percent of households with incomes below the official poverty line were food insecure, compared with 7.0 percent of those with incomes above 185 percent of the poverty line.[9] Differences in food security across demographic and geographic groups reflect, in part, differences in income across those groups, as no adjustment is made for income in the statistics presented in this report.

When Food Insecurity Occurs in U.S. Households, It Is Usually Recurrent but Not Chronic

When households experience very low food security in the United States, the resulting instances of reduced food intake and disrupted eating patterns are usually occasional or episodic but are not usually chronic. The food security measurement methods used in this report are designed to register these occasional or episodic occurrences. The questions used to assess households' food security status ask whether a condition, experience, or behavior occurred at any time in the past 12 months, and households can be classified as having very low food security based on a single, severe episode during the year. It is important to keep this aspect of the scale in mind when interpreting food insecurity statistics. Analyses of additional information collected in the food security survey on how frequently various food-insecure conditions occurred during the year, whether they occurred during the 30 days prior to the survey, and, if so, in how many days, provide insight into the frequency and duration of food insecurity in U.S. households. These analyses reveal that in 2011:

- About one-fourth of the households with very low food security at any time during the year experienced the associated conditions rarely or occasionally—in only 1 or 2 months of the year. For three-fourths of households, the conditions were recurring, experienced in 3 or more months of the year.
- For about one-fourth of food-insecure households and one-third of those with very low food security, occurrence of the associated conditions was frequent or chronic. That is, the conditions occurred often, or in almost every month.
- On average, households that were food insecure at some time during the year were food insecure in 7 months during the year (see Statistical Supplement table S-5, http://www.ers.usda.gov/ publications/ap-administrative-publication/ap-058.aspx). During the 30-day period ending in mid-December 2011, 10.1 million households (8.4 percent of all households) were food insecure—about 56 percent of the number that were food insecure at any time during the year.
- Similarly, households with very low food security at some time during the year experienced the associated conditions, on average, in 7 months during the year (see Statistical Supplement table S-5, http://www.ers.usda.gov/publications/ap-administrative-publication/ap-058.aspx). During the 30-day period ending in mid-December 2011, 4.0 million households (3.4 percent of all households) had very low food security—about 59 percent of the number with very low food security at some during the year.
- Most households that had very low food security at some time during a month experienced the associated conditions in 1 to 7 days of the month. The average daily prevalence of very low food security during the 30-day period ending in mid-December 2011 was probably between 1.0 million and 1.3 million households (0.8 to 1.1 percent of all households)—about 15 to 20 percent of the annual prevalence.
- The daily prevalence of very low food security among children during the 30-day period ending in early December 2011 was probably between 48,000 and 68,000 households (0.12 to 0.17 percent of households with children) about 13 to 18 percent of the annual prevalence.

The omission of homeless families and individuals from these daily statistics biases the statistics downward, and the bias may be substantial relative to the estimates, especially for the most severe conditions.

(Statistical Supplement tables S-8 to S-10 (http://www.ers.usda.gov/publications/ap-administrative-publication/ap-058.aspx) provide information on how often conditions indicating food insecurity occurred, as reported by respondents to the December 2011 food security survey. See Nord et al., 2000, for more information about the frequency of food insecurity.)

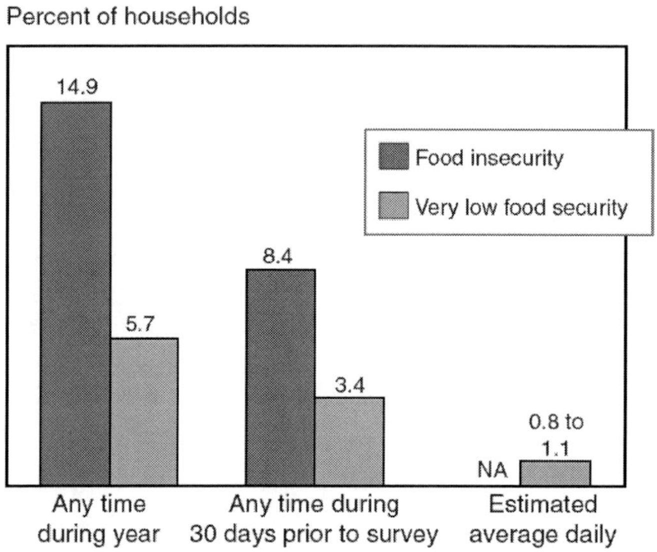

NA = Estimated average daily occurrence of food insecurity is not available because information was not collected on the number of days that less severe food-insecure conditions occurred.

Source: Calculated by USDA, Economic Research Service using data from the December 2011 Current Population Survey Food Security Supplement.

Prevalence of food insecurity and very low food security, by reference period.

Rates of food insecurity were below the national average of 14.9 percent for married-couple families with children (13.9 percent), households with more than one adult and no children (9.9 percent), and households with elderly persons (8.4 percent).[10] The prevalence of food insecurity was also below the national average for White, non-Hispanic households (11.4 percent), households headed by non-Hispanics of other, or multiple, races (12.7 percent), and households with incomes above 185 percent of the poverty line (7.0 percent).

Table 2. Households by food security status and selected household characteristics, 2011

Category	Total[1]	Food secure		Food insecure					
				All		With low food security		With very low food security	
	1,000	*1,000*	*Percent*	*1,000*	*Percent*	*1,000*	*Percent*	*1,000*	*Percent*
All households	119,484	101,631	85.1	17,853	14.9	11,014	9.2	6,839	5.7
Household composition:									
With children < 18 yrs	38,803	30,814	79.4	7,989	20.6	5,750	14.8	2,239	5.8
With children < 6 yrs	17,281	13,494	78.1	3,787	21.9	2,884	16.7	903	5.2
Married-couple families	25,421	21,879	86.1	3,542	13.9	2,671	10.5	871	3.4
Female head, no spouse	9,690	6,124	63.2	3,566	36.8	2,452	25.3	1,114	11.5
Male head, no spouse	3,055	2,295	75.1	760	24.9	532	17.4	228	7.5
Other household with child[2]	638	517	81.0	121	19.0	96	15.0	25	3.9
With no children < 18 yrs	80,681	70,817	87.8	9,864	12.2	5,264	6.5	4,600	5.7
More than one adult	47,863	43,103	90.1	4,760	9.9	2,778	5.8	1,982	4.1
Women living alone	18,021	15,205	84.4	2,816	15.6	1,387	7.7	1,429	7.9
Men living alone	14,796	12,508	84.5	2,288	15.5	1,099	7.4	1,189	8.0
With elderly	30,099	27,579	91.6	2,520	8.4	1,589	5.3	931	3.1
Elderly living alone	11,402	10,400	91.2	1,002	8.8	579	5.1	423	3.7
Race/ethnicity of households:									
White, non-Hispanic	83,304	73,823	88.6	9,481	11.4	5,689	6.8	3,792	4.6
Black, non-Hispanic	14,765	11,066	74.9	3,699	25.1	2,155	14.6	1,544	10.5
Hispanic[3]	14,410	10,629	73.8	3,781	26.2	2,583	17.9	1,198	8.3
Other	7,005	6,113	87.3	892	12.7	587	8.4	305	4.4
Household income-to-poverty ratio:									
Under 1.00	15,557	9,164	58.9	6,393	41.1	3,616	23.2	2,777	17.9
Under 1.30	22,168	13,825	62.4	8,343	37.6	4,763	21.5	3,580	16.1
Under 1.85	31,437	20,593	65.5	10,844	34.5	6,376	20.3	4,468	14.2

Table 2. (Continued)

Category	Total[1]	Food secure		Food insecure					
				All		With low food security		With very low food security	
	1,000	*1,000*	*Percent*	*1,000*	*Percent*	*1,000*	*Percent*	*1,000*	*Percent*
1.85 and over	62,244	57,888	93.0	4,356	7.0	2,938	4.7	1,418	2.3
Income unknown	25,802	23,149	89.7	2,653	10.3	1,700	6.6	953	3.7
Area of residence:[4]									
Inside metropolitan area	99,835	84,999	85.1	14,836	14.9	9,163	9.2	5,673	5.7
In principal cities[5]	33,515	27,597	82.3	5,918	17.7	3,646	10.9	2,272	6.8
Not in principal cities	49,253	42,770	86.8	6,483	13.2	4,013	8.1	2,470	5.0
Outside metropolitan area	19,649	16,632	84.6	3,017	15.4	1,851	9.4	1,166	5.9
Census geographic region:									
Northeast	21,650	18,736	86.5	2,914	13.5	1,777	8.2	1,137	5.3
Midwest	26,359	22,798	86.5	3,561	13.5	2,141	8.1	1,420	5.4
South	44,772	37,618	84.0	7,154	16.0	4,431	9.9	2,723	6.1
West	26,703	22,478	84.2	4,225	15.8	2,665	10.0	1,560	5.8

[1] Totals exclude households for which food security status is unknown because they did not give a valid response to any of the questions in the food security scale. In 2011, these exclusions represented 405,000 households (0.3 percent of all households).

[2] Households with children in complex living arrangements, e.g., children of other relatives or unrelated roommate or boarder.

[3] Hispanics may be of any race.

[4] Metropolitan area residence is based on 2003 Office of Management and Budget delineation.

[5] Households within incorporated areas of the largest cities in each metropolitan area. Residence inside or outside of principal cities is not identified for about 17 percent of households in metropolitan statistical areas.

Source: Calculated by USDA, Economic Research Service using data from the December 2011 Current Population Survey Food Security Supplement.

Rates of food insecurity were higher than the national average for the following groups:

- All households with children (20.6 percent)[11]
- Households with children under age 6 (21.9 percent)
- Households with children headed by a single woman (36.8 percent) or a single man (24.9 percent)[12]
- Black, non-Hispanic households (25.1 percent) and Hispanic households (26.2 percent)
- Low-income households with incomes below 185 percent of the poverty threshold (34.5 percent)

Across the metropolitan area classifications, the prevalence of food insecurity was highest for households located in principal cities of metropolitan areas (17.7 percent), intermediate for those in nonmetropolitan areas (15.4 percent), and lowest in suburbs and other metropolitan areas outside principal cities (13.2 percent).[13] Regionally, the prevalence of food insecurity was higher in the South (16.0 percent) and West (15.8 percent) than in the Midwest (13.5 percent) and Northeast (13.5 percent).

The prevalence of very low food security in various types of households followed a pattern similar to that observed for food insecurity. Percentages were lower than the national average of 5.7 percent for married couples with children (3.4 percent); multiple-adult households with no children (4.1 percent); households with elderly persons (3.1 percent); White, non-Hispanic households (4.6 percent); households headed by non-Hispanics of other races (4.4 percent); households with incomes above 185 percent of the poverty line (2.3 percent); and households residing outside principal cities within metropolitan areas (5.0 percent).

Very low food security was more prevalent than the national average (5.7 percent) for the following groups:

- Households with children headed by a single woman (11.5 percent) or a single man (7.5 percent)
- Women living alone (7.9 percent) and men living alone (8.0 percent)
- Black, non-Hispanic households (10.5 percent) and Hispanic households (8.3 percent)
- Households with incomes below 185 percent of the poverty line (14.2 percent)

- Households located in principal cities of metropolitan areas (6.8 percent).

In 10.0 percent of households with children, one or more child was food insecure (table 3).[14] The percentage of households with food-insecure children was lowest in married-couple households (6.3 percent); White, non-Hispanic households (6.7 percent); households with incomes above 185 percent of the poverty line (3.1 percent); metropolitan households located outside of principal cities (8.8 percent); and households in the Northeast (8.5 percent) and Midwest (8.0 percent). The percentage of households with food-insecure children was highest for female-headed households (18.9 percent); male-headed households (12.1 percent); Black, non-Hispanic households (14.6 percent); Hispanic households (17.4 percent); households with incomes below 185 percent of the poverty line (20.8 percent); households within principal cities of metropolitan areas (12.0 percent); and households in the West (11.6 percent).

Very low food security among children was least prevalent in married-couple households; White, non-Hispanic households; and households with incomes above 185 percent of the poverty line. Very low food security among children was most prevalent in households headed by a single woman (1.8 percent); households headed by a Black, non-Hispanic person (1.9 percent) or Hispanic person (1.9 percent); and households with incomes below 185 percent of the poverty line (2.2 percent).

Prevalence rates of food insecurity remained essentially unchanged from 2010 to 2011 in most categories analyzed (figure 4). That is, the differences in the estimates could have resulted from sampling variation. However, the prevalence of food insecurity increased among women living alone and households headed by non-Hispanic Whites.

The prevalence of very low food insecurity increased in some household categories from 2010 to 2011 and remained essentially unchanged in the rest (figure 5). Increases were statistically significant for women living alone; households with elderly members; White, non-Hispanic households; Black, non-Hispanic households; and households with incomes below 185 percent of the poverty line.

Table 3. Prevalence of food security and food insecurity in households with children by selected household characteristics, 2011

Category	Total[1]	Food-secure households		Food-insecure households[2]		Households with food-insecure children[3]		Households with very low food security among children	
	1,000	*1,000*	*Percent*	*1,000*	*Percent*	*1,000*	*Percent*	*1,000*	*Percent*
All households with children	38,803	30,814	79.4	7,989	20.6	3,862	10.0	374	1.0
Household composition:									
With children < 6 yrs	17,281	13,494	78.1	3,787	21.9	1,668	9.7	123	.7
Married-couple families	25,421	21,879	86.1	3,542	13.9	1,612	6.3	164	.6
Female head, no spouse	9,690	6,124	63.2	3,566	36.8	1,832	18.9	177	1.8
Male head, no spouse	3,055	2,296	75.2	759	24.8	371	12.1	NA	NA
Other household with child4	638	517	81.0	121	19.0	46	7.2	NA	NA
Race/ethnicity of households:									
White, non-Hispanic	23,274	19,655	84.5	3,619	15.5	1,552	6.7	121	.5
Black, non-Hispanic	5,284	3,742	70.8	1,542	29.2	769	14.6	98	1.9
Hispanic[5]	7,447	5,045	67.7	2,402	32.3	1,294	17.4	138	1.9
Other	2,799	2,373	84.8	426	15.2	247	8.8	NA	NA
Household income-to-poverty ratio:									
Under 1.00	7,090	3,882	54.8	3,208	45.2	1,711	24.1	199	2.8
Under 1.30	9,278	5,240	56.5	4,038	43.5	2,122	22.9	251	2.7
Under 1.85	13,086	7,816	59.7	5,270	40.3	2,724	20.8	290	2.2
1.85 and over	18,986	17,469	92.0	1,517	8.0	588	3.1	36	.2
Income unknown	6,731	5,530	82.2	1,201	17.8	549	8.2	47	.7
Area of residence:[5]									
Inside metropolitan area	32,915	26,224	79.7	6,691	20.3	3,248	9.9	327	1.0
In principal cities[7]	10,400	7,883	75.8	2,517	24.2	1,247	12.0	129	1.2

Table 3. (Continued)

Category	Total[1]	Food-secure households		Food-insecure households[2]		Households with food-insecure children[3]		Households with very low food security among children	
	1,000	*1,000*	*Percent*	*1,000*	*Percent*	*1,000*	*Percent*	*1,000*	*Percent*
Not in principal cities	17,066	14,004	82.1	3,062	17.9	1,502	8.8	154	.9
Outside metropolitan area	5,888	4,590	78.0	1,298	22.0	613	10.4	46	.8
Census geographic region:									
Northeast	6,797	5,538	81.5	1,259	18.5	580	8.5	56	.8
Midwest	8,262	6,835	82.7	1,427	17.3	657	8.0	66	.8
South	14,454	11,289	78.1	3,165	21.9	1,549	10.7	189	1.3
West	9,290	7,153	77.0	2,137	23.0	1,076	11.6	63	0.7

NA = Not reported; fewer than 10 households in the survey with this characteristic had very low food security among children.

[1] Totals exclude households for which food security status is unknown because they did not give a valid response to any of the questions in the food security scale. In 2011, these exclusions represented 154,000 households with children (0.4 percent of all households with children).

[2] Food-insecure households are those with low or very low food security among adults or children or both.

[3] Households with food-insecure children are those with low or very low food security among children.

[4] Households with children in complex living arrangements, e.g., children of other relatives or unrelated roommate or boarder.

[5] Hispanics may be of any race.

[6] Metropolitan area residence is based on 2003 Office of Management and Budget delineation.

[7] Households within incorporated areas of the largest cities in each metropolitan area. Residence inside or outside of principal cities is not identified for about 17 percent of households in metropolitan statistical areas.

Source: Calculated by USDA, Economic Research Service using data from the December 2011 Current Population Survey Food Security Supplement.

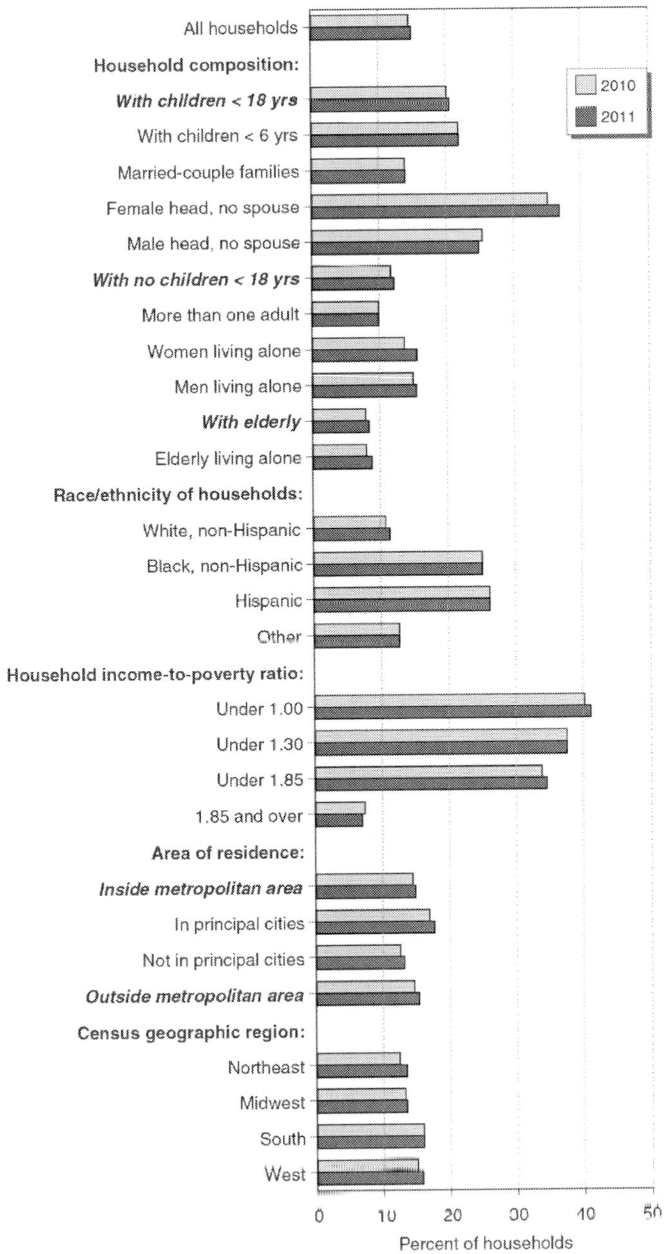

Source: Calculated by USDA, Economic Research Service based on Current Population Survey Food Security Supplement data.

Figure 4. Prevalence of food insecurity, 2010 and 2011.

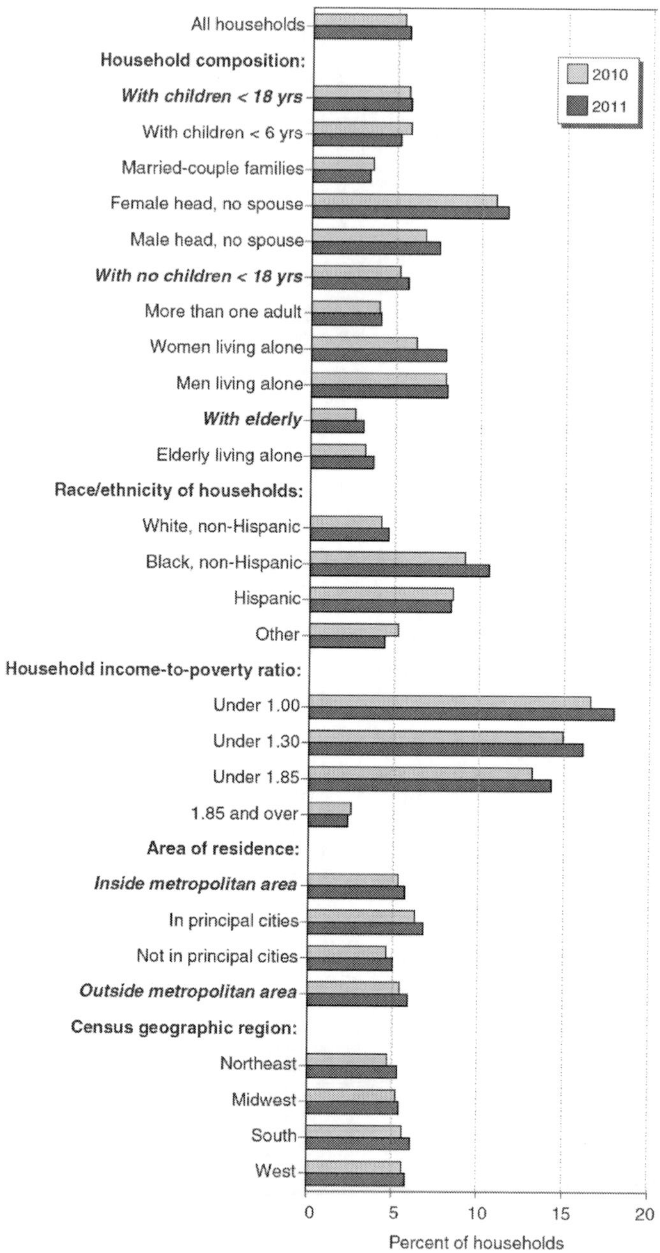

Source: Calculated by USDA, Economic Research Service based on Current Population Survey Food Security Supplement data.

Figure 5. Prevalence of very low food security, 2010 and 2011.

Number of Persons by Household Food Security Status

The food security survey is designed to measure food security status at the household level. While it is informative to examine the number of persons residing in food-insecure households, these statistics should be interpreted carefully. Within a food-insecure household, each household member may be affected differently by the household's food insecurity. Some members—particularly young children—may experience only mild effects or none at all, while adults are more severely affected. It is more precise, therefore, to describe these statistics as representing "persons living in food-insecure households" rather than as representing "food-insecure persons." Similarly, "persons living in households with very low food security" is a more precise description than "persons with very low food security."

In 2011, 50.1 million people lived in food-insecure households (see table 1A). They constituted 16.4 percent of the U.S. civilian noninstitutionalized population and included 33.5 million adults and 16.7 million children (see table 1B). About 8.6 million children (11.5 percent) lived in households in which one or more child was food insecure. About 12.1 million adults (5.2 percent) lived in households with very low food security (see table 1A), and 845,000 children (1.1 percent) lived in households with very low food security among children (see table 1B).

Statistical Supplement tables S-2 and S-3 present estimates of the number of people and the number of children in households in each food security status and household type (http://www.ers.usda.gov/publications/ap-administrative-publication/ap-058.aspx).

Prevalence of Food Insecurity by State

The prevalence of food insecurity varied considerably from State to State. Data from 3 years, 2009-11, were combined to provide more reliable statistics at the State level (table 4). Estimated prevalence of food insecurity during this 3-year period ranged from 7.8 percent in North Dakota to 19.2 percent in Arkansas and Mississippi; estimated prevalence rates of very low food security ranged from 3.1 percent in North Dakota to 7.6 percent in Arkansas.

The margins of error for the State prevalence rates should be taken into consideration when interpreting these statistics and especially when comparing prevalence rates across States. The margin of error reflects sampling variation—the uncertainty associated with estimates that are based on information from a limited number of households in each State. The margins of error presented in table 4 indicate the range (above or below the estimated prevalence rate) that is 90 percent likely to include the true prevalence rate. For example, considering the margins of error, it is not certain that the prevalence of very low food security was higher in Arkansas than in the States with the next 16 highest prevalence rates.

Table 4. Prevalence of household-level food insecurity and very low food security by State, average 2009-11[1]

States	Number of households		Food insecurity (low or very low food security)		Very low food security	
	Average 2009-11[2]	Interviewed	Prevalence	Margin of error[3]	Prevalence	Margin of error[3]
	Number	*Number*	*Percent*	*Percentage points*	*Percent*	*Percentage points*
U.S.	118,805,000	133,589	14.7	0.25	5.6	0.20
AK	257,000	1,736	13.9	1.85	4.8	1.10
AL	1,798,000	1,519	18.2*	2.11	7.1*	1.40
AR	1,186,000	1,527	19.2*	2.13	7.6*	1.39
AZ	2,575,000	1,621	15.8	1.45	6.2	1.19
CA	12,992,000	9,988	16.2*	.76	5.8	.43
CO	2,054,000	2,885	13.4	1.57	5.4	.65
CT	1,349,000	2,893	11.9*	.82	4.7	.99
DC	297,000	2,103	12.6*	1.70	4.4*	1.17
DE	347,000	2,003	10.9*	1.29	4.3*	.78
FL	7,627,000	5,130	15.4	.68	6.3*	.59
GA	3,786,000	2,787	17.4*	2.00	6.4	1.07
HI	443,000	1,802	13.8	1.19	5.6	.63
IA	1,236,000	2,557	12.0*	1.04	4.7*	.83
ID	578,000	1,551	13.7	1.48	5.0	1.09
IL	4,941,000	4,120	13.3*	1.00	4.5*	.59
IN	2,515,000	2,137	13.2	1.93	5.4	1.13
KS	1,151,000	2,112	14.5	1.75	5.2	1.26
KY	1,789,000	2,129	16.4	1.65	6.6*	.94
LA	1,819,000	1,371	14.1	1.55	4.4*	1.16
MA	2,570,000	2,033	11.9*	1.24	4.5	1.19
MD	2,229,000	2,988	12.5*	1.46	5.6	.94
ME	542,000	2,668	14.7	1.53	6.5	.96
MI	3,855,000	3,207	14.2	1.15	5.6	.93
MN	2,148,000	3,421	10.2*	.68	4.3*	.44
MO	2,436,000	2,220	16.0	1.27	6.7*	.77
MS	1,105,000	1,285	19.2*	1.81	7.1*	1.41
MT	424,000	1,460	14.5	1.74	5.4	.96
NC	3,720,000	2,862	17.1*	1.11	5.8	.73

States	Number of households		Food insecurity (low or very low food security)		Very low food security	
	Average 2009-11[2]	Interviewed	Prevalence	Margin of error[3]	Prevalence	Margin of error[3]
	Number	*Number*	*Percent*	*Percentage points*	*Percent*	*Percentage points*
ND	274,000	1,901	7.8*	1.30	3.1*	.68
NE	718,000	2,045	13.3	1.92	4.9	.87
NH	524,000	2,929	9.6*	1.01	4.0*	.74
NJ	3,182,000	2,497	12.3*	1.07	4.5*	.67
NM	807,000	1,039	16.5	1.99	6.3	1.35
NV	1,002,000	1,971	15.3	1.84	6.0	1.37
NY	7,615,000	5,569	13.3*	.62	5.1*	.42
OH	4,577,000	3,821	15.5	1.50	6.4	.85
OK	1,481,000	1,651	14.7	1.35	7.0*	1.03
OR	1,496,000	2,011	13.6	1.46	5.9	1.00
PA	5,116,000	4,055	12.5*	.86	5.0*	.45
RI	432,000	2,386	15.5	1.48	6.0	1.02
SC	1,809,000	1,978	14.8	1.92	4.4*	.69
SD	332,000	2,209	12.7*	.75	5.3	.59
TN	2,519,000	1,944	15.2	1.19	6.4*	.77
TX	9,054,000	6,491	18.5*	.75	6.5*	.46
UT	944,000	1,413	14.6	1.58	5.2	.76
VA	2,978,000	2,810	9.1*	1.19	3.2*	.56
VT	265,000	2,120	12.8*	1.59	5.4	1.16
WA	2,629,000	2,431	15.4	1.44	6.2	1.19
WI	2,303,000	2,789	11.3*	1.01	4.7*	.70
WV	751,000	1,626	14.1	1.97	5.1	1.05
WY	227,000	1,788	12.7*	1.96	4.8	1.07

* Difference from U.S. average was statistically significant with 90-percent confidence ($t > 1.645$).

[1] Prevalence rates for 1996-98 reported in *Prevalence of Food Insecurity and Hunger, by State, 1996-1998* (Nord et al., 1999) are not directly comparable with the rates reported here because of differences in screening procedures in the CPS Food Security Supplements from 1995 to 1998. Comparable statistics for 1996-1998 are presented in *Food Security in the United States in 2010 Statistical Supplement* table S4 (Coleman-Jensen et al., 2011).

[2] Totals exclude households for which food security status is unknown because they did not give a valid response to any of the questions in the food security scale. These exclusions represented about 0.3 percent of all households in 2009, 0.4 percent in 2010, and 0.3 percent in 2011.

[3] Margin of error with 90-percent confidence (1.645 times the standard error of the estimated prevalence rate)

Source: Prepared by USDA, Economic Research Service using Current Population Survey Food Security Supplement data.

Taking into account margins of error of the State and U.S. estimates, the prevalence of food insecurity was higher (i.e., statistically significantly higher) than the national average in 7 States and lower than the national average in 17 States and the District of Columbia. In the remaining 26 States, differences from the national average were not statistically significant. The prevalence of very low food security was higher than the national average in 9 States, lower than the national average in 13 States and the District of Columbia, and not significantly different from the national average in 28 States.

State-level prevalence rates of food insecurity and very low food security for the period 2009-11 are compared with 3-year average rates for 2006-08 and 1999-2001 in table S-4 of the Statistical Supplement tables (http://www.ers.usda.gov/publications/ap-administrative-publication/ap-058.aspx/).

Household Spending on Food

This section provides information on how much households spent on food, as reported in the December 2011 food security survey. Food insecurity is a condition that arises from lack of money and other resources to acquire food. In most households, the majority of food consumed by household members is purchased—either from supermarkets or grocery stores—to be prepared and eaten at home, or from cafeterias, restaurants, or vending machines to be eaten outside the home. The amount of money that a household spends on food, therefore, provides insight into how adequately the household is meeting its food needs.[15] When a household reduces food spending below some minimum level because of constrained resources, various aspects of food insecurity such as disrupted eating patterns and reduced food intake may result.

Methods

The household food expenditure statistics in this report are based on usual weekly spending for food, as reported by respondents after they were given a chance to reflect on the household's actual food spending during the previous week. Respondents were first asked to report the amounts of money their households had spent on food in the week prior to the interview, including any purchases made with SNAP benefits (formerly called food stamps) at:

- supermarkets and grocery stores;
- stores other than supermarkets and grocery stores such as meat markets, produce stands, bakeries, warehouse clubs, and convenience stores;
- restaurants, fast food places, cafeterias, and vending machines;
- "...any other kind of place."[16]

Total spending for food, based on responses to this series of questions, was verified with the respondent, and the respondent was then asked how much the household usually spent on food during a week. Analyses by ERS researchers have found that usual food expenditures estimated from data collected by this method were consistent with estimates from the Consumer Expenditure Survey (CES)—the principal source of data on U.S. household expenditures for goods and services (Oliveira and Rose, 1996; Nord 2009b).

Food spending was adjusted for household size and composition in two ways. The first adjustment was calculated by dividing each household's usual weekly food spending by the number of persons in the household, yielding the "usual weekly food spending per person" for that household. The second adjustment accounts more precisely for the different food needs of households by comparing each household's usual food spending to the estimated cost of the Thrifty Food Plan for that household in December 2011.[17] The Thrifty Food Plan—developed by USDA—serves as a national standard for a nutritious, low-cost diet. It represents a set of "market baskets" of food that people in specific age and gender categories could consume at home to maintain a healthful diet that meets current dietary standards, taking into account the food consumption patterns of U.S. households (U.S. Department of Agriculture, Center for Nutrition Policy and Promotion, 2007).[18] Each household's reported usual weekly food spending was divided by the household-specific cost of the Thrifty Food Plan based on the age and gender of each household member and the number of persons in the household (U.S. Department of Agriculture, Center for Nutrition Policy and Promotion, 2012).[19]

The medians of each of the two food spending measures (spending per person per week and spending relative to the cost of the Thrifty Food Plan) were estimated at the national level and for households in various categories to represent the usual weekly food spending of the typical household in each category. Medians are reported rather than averages (means) because medians are not unduly affected by the few unexpectedly high values of usual food spending that are believed to be reporting errors or data entry errors. Thus, the median better reflects what a typical household spent.

About 6.4 percent of households interviewed in the CPS food security survey did not respond to the food spending questions or reported zero usual food spending and were excluded from the analysis. As a result, the total number of households represented in tables 5 and 6 is somewhat smaller than that in tables 1 and 2, and food spending estimates may not be fully representative of all households in the United States.[20]

Food Expenditures, by Selected Household Characteristics

In 2011, the typical U.S. household spent $47.50 per person each week for food (table 5). Median household food spending relative to the cost of the Thrifty Food Plan—which adjusts for food price inflation and adjusts more precisely for the food needs of persons in different age-gender categories—was 1.15, down from 1.17 in 2010. That is, in 2011, the typical household spent 15 percent more on food than the cost of the Thrifty Food Plan for that household.

Table 5. Weekly household food spending per person and relative to the cost of the Thrifty Food Plan (TFP), 2011

Category	Number of households[1]	Median weekly food spending	
		Per person	Relative to cost of TFP
	1,000	*Dollars*	*Ratio*
All households	111,305	47.50	1.15
Household composition:			
With children < 18 yrs	36,833	35.00	1.00
At least one child < 6 yrs	16,451	32.50	1.00
Married-couple families	24,199	36.67	1.03
Female head, no spouse	9,207	33.33	.94
Male head, no spouse	2,848	35.00	.98
Other household with child[2]	579	32.50	.88
With no children < 18 yrs	74,473	50.00	1.22
More than one adult	44,400	50.00	1.17
Women living alone	16,367	55.00	1.26
Men living alone	13,705	70.00	1.43

Category	Number of households[1]	Median weekly food spending	
		Per person	Relative to cost of TFP
	1,000	*Dollars*	*Ratio*
With elderly	27,049	50.00	1.14
Elderly living alone	9,967	50.00	1.17
Race/ethnicity of households:			
White, non-Hispanic	77,792	50.00	1.21
Black, non-Hispanic	13,520	40.00	.99
Hispanic[3]	13,500	37.50	.98
Other	6,493	43.75	1.10
Household income-to-poverty ratio:			
Under 1.00	14,713	35.00	.93
Under 1.30	21,024	35.00	.93
Under 1.85	29,853	36.67	.93
1.85 and over	59,792	50.00	1.26
Income unknown	21,661	45.83	1.13
Area of residence:[4]			
Inside metropolitan area	92,999	50.00	1.17
In principal cities[5]	31,172	50.00	1.16
Not in principal cities	45,851	50.00	1.17
Outside metropolitan area	18,306	40.00	1.04
Census geographic region:			
Northeast	20,186	50.00	1.21
Midwest	24,370	43.33	1.10
South	41,742	45.67	1.14
West	25,007	50.00	1.18

[1] Totals exclude households that did not answer the questions about spending on food or reported zero usual food spending. These exclusions represented 7.2 percent of all households.

[2] Households with children in complex living arrangements, e.g., children of other relatives or unrelated roommate or boarder.

[3] Hispanics may be of any race.

[4] Metropolitan area residence is based on 2003 Office of Management and Budget delineation.

[5] Households within incorporated areas of the largest cities in each metropolitan area. Residence inside or outside of principal cities is not identified for about 17 percent of households in metropolitan statistical areas.

Source: Calculated by USDA, Economic Research Service using data from the December 2011 Current Population Survey Food Security Supplement.

Households with children under age 18 generally spent less for food, relative to the cost of the Thrifty Food Plan, than those without children. The typical household with children spent an amount on food equivalent to the cost of the Thrifty Food Plan, while the typical household with no children spent 22 percent more than the cost of the Thrifty Food Plan. Median food expenditures relative to the cost of the Thrifty Food Plan were lower for households with children headed by single women (0.94) than for married couples with children (1.03). Median food expenditure relative to the cost of the Thrifty Food Plan was highest for men living alone (1.43).

Median food expenditures relative to the cost of the Thrifty Food Plan were lower for Black, non-Hispanic households (0.99) and Hispanic households (0.98) than for White, non-Hispanic households (1.21). This pattern is consistent with the lower average incomes and higher prevalence rates of food insecurity of these racial and ethnic minorities.

As expected, households with higher incomes spent more money on food than lower income households.[21] The typical household with income below the poverty line spent about 7 percent less than the cost of the Thrifty Food Plan, while the typical household with income above 185 percent of the poverty line spent 26 percent more than the cost of the Thrifty Food Plan.

Table 6. Weekly household food spending per person and relative to the cost of the Thrifty Food Plan (TFP) by food security status, 2011

Category	Number of households[1]	Median weekly food spending	
		Per person	Relative to cost of TFP
	1,000	*Dollars*	*Ratio*
All households	111,305	47.50	1.15
Food security status:			
Food-secure households	94,061	50.00	1.17
Food-insecure households	17,050	37.00	.94
Households with low food security	10,545	35.71	.94
Households with very low food security	6,505	37.50	.93

[1] Total for all households excludes households that did not answer the questions about spending on food or reported zero usual spending for food. These represented 7.2 percent of all households. Totals in the bottom section also exclude households that did not answer any of the questions in the food security scale.

Source: Calculated by USDA, Economic Research Service using data from the December 2011 Current Population Survey Food Security Supplement.

Median food spending relative to the cost of the Thrifty Food Plan was lower for households outside metropolitan areas (1.04) than for those inside metropolitan statistical areas (1.17). Regionally, median spending on food relative to the cost of the Thrifty Food Plan was lowest in the Midwest (1.10) and highest in the Northeast (1.21).

Food Expenditures and Household Food Security

Food-secure households typically spent more on food than food-insecure households. Median food spending relative to the cost of the Thrifty Food Plan was 1.17 among food-secure households, compared with 0.94 among food-insecure households (table 6). Taking into account estimated food need, food-secure households spent approximately 24 percent more for food than food-insecure households.

Federal Food and Nutrition Assistance Programs and Food Security

Households with limited resources employ a variety of methods to help meet their food needs. Some participate in one or more of the Federal food and nutrition assistance programs or obtain food from emergency food providers in their communities to supplement the food they purchase. Households that turn to Federal and community food and nutrition assistance programs typically do so because they are having difficulty in meeting their food needs. The use of such programs by low-income households and the relationship between their food security status and use of food and nutrition assistance programs provide insight into the extent of these households' difficulties in obtaining enough food and the ways they cope with those difficulties.

This section presents information about the food security status of households that participated in the three largest Federal food and nutrition assistance programs (see box, "Federal Food and Nutrition Assistance Programs," on page 24). It also provides information about the extent to which food-insecure households participated in these programs. Total participation in the Federal food and nutrition assistance programs, participation rates of eligible households in those programs, and characteristics of participants in those programs are not described in this report. Extensive information on those topics is available from USDA's Food and Nutrition Service (FNS).[22]

Statistical Supplement tables S-12 to S-17 provide information on food spending by participants and low-income nonparticipants in selected Federal and community food and nutrition assistance programs and about the extent to which households obtained assistance from community food pantries and emergency kitchens (http://www.ers.usda.gov/publications/ap-administrative-publication/ap-058.aspx).

Methods

The December 2011 CPS food security survey included questions about the use of Federal food and nutrition assistance programs. All households with incomes below 185 percent of the Federal poverty threshold were asked these questions. In order to minimize the burden on respondents, households with incomes above that range were not asked the questions unless they indicated some level of difficulty in meeting their food needs on the first of the preliminary screener questions listed in footnote 5. The questions analyzed in this section are:

- "During the past 12 months…did anyone in this household get SNAP or food stamp benefits?"[23] Households that responded affirmatively were then asked in which months they received SNAP benefits and on what date they last received them. Information from these 3 questions was combined to identify households that received SNAP benefits in the 30 days prior to the survey.
- "During the past 30 days, did any children in the household…receive free or reduced-cost lunches at school?" (Only households with children between the ages of 5 and 18 were asked this question.)
- "During the past 30 days, did any women or children in this household get food through the WIC program?" (Only households with a child age 0-4 or a woman age 15-45 were asked this question.)

Prevalence rates of food security, food insecurity, and very low food security were calculated for households reporting use of each food and nutrition assistance provider and for comparison groups of nonparticipating households with incomes and household compositions similar to those of food assistance recipients. Statistics for participating households excluded households with incomes above the ranges specified for the comparison groups.[24] The proportions of food-insecure households participating in each of the three largest Federal food and nutrition assistance programs—the Supplemental Nutrition Assistance Program; the National School Lunch Program; and the Special Supplemental Nutrition Program for Women, Infants, and Children—were calculated, as well as the proportion that participated in any of the three programs. These analyses were restricted to households with annual incomes below 185 percent of the poverty line because most households with incomes above this range were not asked whether they participated in these programs.

FEDERAL FOOD AND NUTRITION ASSISTANCE PROGRAMS

USDA's Food and Nutrition Service (FNS) administers 15 domestic food and nutrition assistance programs. The three largest programs are (for more information, see Oliveira, 2012):

- The Supplemental Nutrition Assistance Program (SNAP), formerly called the Food Stamp Program. The program provides monthly benefits for eligible low-income households to purchase approved food items at authorized food stores. Participants qualify for the program based on available household income, assets, and certain basic expenses. In an average month of fiscal year 2011 (October 1, 2010 through September 30, 2011), SNAP provided benefits to 44.7 million people in the United States (14 percent of individuals). The average benefit was about $134 per person per month, and total Federal expenditures for the program were over $75 billion.
- The National School Lunch Program. The program operates in over 100,000 public and nonprofit private schools and residential child-care institutions. All meals served under the program receive Federal subsidies, and free or reduced-price lunches are available to low-income students. In fiscal year 2011, the program provided lunches to an average of 31.8 million children each school day. Fifty-eight percent of the lunches served in 2011 were free, and an additional 8 percent were provided at reduced prices.

- The Special Supplemental Nutrition Program for Women, Infants, and Children (WIC). The program is a federally funded preventive nutrition program that provides grants to States to support distribution of supplemental foods, health care referrals, and nutrition education for low-income pregnant, breastfeeding, and nonbreastfeeding postpartum women, for infants in low-income families, and for children in low-income families who are younger than 5 years old and who are found to be at nutritional risk. Most State WIC programs provide vouchers that participants use to acquire supplemental food packages at authorized food stores. In fiscal year 2011, WIC served an average 9 million participants per month at an average monthly cost for food (after rebates to the WIC program from manufacturers) of about $47 per person.

Food Security of Households That Received Food and Nutrition Assistance

The relationship between food security and the use of food and nutrition assistance programs is complex. There are reasons to expect that households that report using food and nutrition assistance programs in a one-time survey can either be more food secure or less food secure than low-income households not using those programs. Since the programs provide food and other resources to reduce the severity of food insecurity, households are expected to be more food secure after receiving program benefits than before doing so. On the other hand, it is the more food-insecure households, those having greater difficulty meeting their food needs, that seek assistance from the programs.[25] About 52 percent of households that received SNAP benefits were food insecure, as were 48.7 percent of households that received free or reduced-cost school lunches, and 43.7 percent of those that received WIC benefits (table 7). The prevalence of very low food security among households participating in SNAP was 11.7 percentage points higher than that of nonparticipating households in the same low-income range (23.0 percent versus 11.3 percent). For households that received free or reduced-cost school lunches, the prevalence of very low food security was more than twice that of nonparticipating households with school-age children in the same income range (16.6 percent versus 7.9 percent).

A possible complicating factor in interpreting table 7 is that food insecurity was measured over a 12-month period. An episode of food insecurity may have occurred at a different time during the year than the use of a specific food and nutrition assistance program. A similar tabulation using a 30-day measure of food insecurity largely overcomes this potential problem because measured food insecurity and reported use of food and nutrition assistance programs are more likely to refer to contemporaneous conditions when both are referenced to the previous 30 days. That tabulation shows patterns of food insecurity and the use of food and nutrition assistance programs that are similar to those in table 7, although 30-day prevalence rates were somewhat lower than the corresponding 12-month rates (see Statistical Supplement table S-16, http://www.ers.usda.gov/publications/ap-administrative-publication/ap-058.aspx).

Table 7. Percent of households by food security status and participation in selected Federal food and nutrition assistance programs, 2011

Category	Food secure	Food insecure		
		All	With low food security	With very low food security
	Percent			
Income less than 130 percent of poverty line:				
Received SNAP[1] benefits previous 12 months	48.3	51.7	28.7	23.0
Received SNAP benefits all 12 months	50.9	49.1	26.8	22.3
Received SNAP benefits 1 to 11 months	44.0	56.0	31.8	24.2
Did not receive SNAP benefits previous 12 months	72.3	27.7	16.4	11.3
Income less than 185 percent of poverty line; school-age children in household:				
Received free or reduced-price school lunch previous 30 days	51.3	48.7	32.1	16.6
Did not receive free or reduced-price school lunch previous 30 days	72.3	27.7	19.8	7.9
Income less than 185 percent of poverty line; children under age 5 in household:				
Received WIC[2] previous 30 days	56.3	43.7	30.1	13.6
Did not receive WIC previous 30 days	66.7	33.3	25.8	7.5

[1] SNAP = Supplemental Nutrition Assistance Program, formerly the Food Stamp Program.

[2] WIC = Special Supplemental Nutrition Program for Women, Infants, and Children.

Source: Calculated by USDA, Economic Research Service using data from the December 2011 Current Population Survey Food Security Supplement.

Participation in Federal Food and Nutrition Assistance Programs by Food-Insecure Households

About 57 percent of food-insecure households reported receiving assistance from one or more of the three largest Federal food and nutrition assistance programs during the month prior to the December 2011 food security survey (table 8). SNAP provided assistance to 40.1 percent of food-insecure households, children in 32.2 percent received free or reduced-price school lunches, and women or children in 11.2 percent received WIC food vouchers.[26] Fifty-six percent of households classified as having very low food security reported participating in one or more of the three largest Federal food and nutrition assistance programs, and the largest share of these (42.5 percent) participated in SNAP.[27]

Table 8. Participation of food-insecure households in selected Federal food and nutrition assistance programs, 2011

Program	Share of food-insecure households that participated in the program during the previous 30 days[1,2]	Share of households with very low food security that participated in the program during the previous 30 days[1,2]
	Percent	
SNAP[3]	40.1	42.5
Free or reduced-price school lunch	32.2	26.5
WIC[4]	11.2	8.6
Any of the three programs	57.2	56.0
None of the three programs	42.8	44.0

[1] Analysis is restricted to households with annual incomes less than 185 percent of the poverty line because most households with incomes above that range were not asked whether they participated in food assistance programs.

[2] These statistics understate the extent of food and nutrition program participation because program participation is underreported by household survey respondents, see footnote 26.

[3] SNAP = Supplemental Nutrition Assistance Program, formerly the Food Stamp Program.

[4] WIC = Special Supplemental Nutrition Program for Women, Infants, and Children.

Source: Calculated by USDA, Economic Research Service using data from the December 2011 Current Population Survey Food Security Supplement.

REFERENCES

Anderson, S. A. (ed.). (1990). "Core Indicators of Nutritional State for Difficult-To-Sample Populations," *Journal of Nutrition*, *120*(11S), 1557-1600. Report by the Life Sciences Research Office, Federation of American Societies for Experimental Biology, for the American Institute of Nutrition.

Andrews, Margaret, Gary Bickel & Steven Carlson. (1998). "Household Food Security in the United States in 1995: Results From the Food Security Measurement Project," *Family Economics and Nutrition Review*, *11*(1&2), 17-28, USDA, Center for Nutrition Policy and Promotion.

Andrews, Margaret, Mark Nord, Gary Bickel, and Steven Carlson. (2000). *Household Food Security in the United States, 1999*. FANRR-8, USDA, Economic Research Service. Available at: http://webarchives. cdlib.org/ sw1m04028n/http://www.ers.usda.gov/publications/fanrr8/.

Bickel, G., Andrews, M. & Carlson, S. (1998). "The Magnitude of Hunger: A New National Measure of Food Security," *Topics in Clinical Nutrition*, *13*(4), 15-30.

Bickel, G., Nord, M., Price, C., Hamilton, W. L. & Cook, J. T. (2000). *Guide to Measuring Household Food Security, Revised 2000*. USDA, Food and Nutrition Service. Available at: http://www.fns.usda.gov/fsec/ files/ fsguide.pdf/.

Carlson, S. J., Andrews, M. S. & Bickel, G.W. (1999). "Measuring Food Insecurity and Hunger in the United States: Development of a National Benchmark Measure and Prevalence Estimates," *The Journal of Nutrition*, *129*, 510S-516S.

Cohen, Barbara, Mark Nord, Robert Lerner, James Parry & Kenneth Yang. (2002). *Household Food Security in the United States, 1998 and 1999: Technical Report.* E-FAN-02-010, prepared by IQ Solutions and USDA, Economic Research Service. Available at: http:// webarchives.cdlib.org/ sw1m04028n/http://www.ers.usda.gov/publications/efan02010/.

Coleman-Jensen, Alisha, Mark Nord, Margaret Andrews, and Steven Carlson. (2011). *Statistical Supplement to Household Food Security in the United States in 2010*. AP-057, USDA, Economic Research Service. Available at: http://www.ers.usda.gov/publications/ap-administrative-publication/ap-057.aspx/.

Gundersen, Craig & Joseph Gruber. (2001). "The Dynamic Determinants of Food Insecurity," Margaret Andrews and Mark Prell (eds.), *Second Food Security Measurement and Research Conference, Volume II: Papers.*

FANRR-11-2, pp. 92-110. USDA, Economic Research Service. Available at: http://webarchives.cdlib.org/sw15d8pg 7m/http://ers.usda.gov/ publications/fanrr11-2/.

Gundersen, Craig & Victor Oliveira. (2001). "The Food Stamp Program and Food Insufficiency," *American Journal of Agricultural Economics*, *83*(4), 875-87.

Hamilton, W. L., Cook, J. T., Thompson, W. W. Buron, L. F., Frongillo, Jr., E. A., Olson, C. M. & Wehler, C. A. (1997a). *Household Food Security in the United States in 1995: Summary Report of the Food Security Measurement Project.* Prepared for USDA, Food and Consumer Service. Available at: http://www.fns.usda.gov/ora/menu/published/ foodsecurity/ sumrpt.pdf/.

Hamilton, W. L., Cook, J. T., Thompson, W. W., Buron, L. F., Frongillo, Jr., E. A., Olson, C. M. & Wehler, C. A. (1997b). *Household Food Security in the United States in 1995: Technical Report.* Prepared for USDA, Food and Consumer Service. Available at: http://www.fns. usda.gov/ora/menu/ published/foodsecurity/tech_rpt.pdf/.

Meyer, Bruce D., Mok, Wallace K. C. & Sullivan, James X. (2009). *The Under-Reporting of Transfers in Household Surveys: Its Nature and Consequences,* NBER Working Paper No. 15181, Cambridge, MA: National Bureau of Economic Research.

National Research Council. (2006). *Food Insecurity and Hunger in the United States: An Assessment of the Measure.* Committee on National Statistics, Panel to Review the U.S. Department of Agriculture's Measurement of Food Insecurity and Hunger, Gooloo S. Wunderlich and Janet L. Norwood (eds.). Washington, DC: The National Academies Press.

Nelson, K., Brown, M. & Lurie, N. (1998). "Hunger in an Adult Patient Population," *Journal of the American Medical Association*, *279*, 1211-14.

Nord, Mark. (2009a). *Food Insecurity in Households with Children: Prevalence, Severity, and Household Characteristics.* EIB-56, USDA, Economic Research Service. Available at: http://www.ers.usda.gov/ publications/eib-economic-information-bulletin/eib56.aspx/.

Nord, Mark. (2009b). *Food Spending Declined and Food Insecurity Increased for Middle-Income and Low-Income Households from 2000 to 2007.* EIB-61, USDA, Economic Research Service. Available at: http:// www.ers. usda.gov/publications/eib-economic-information-bulletin/ eib61.aspx/.

Nord, Mark, Margaret Andrews & Steven Carlson. (2002). *Household Food Security in the United States, 2001.* FANRR-29, USDA, Economic

Research Service. Available at: http://webarchives.cdlib.org/sw15d8pg7m/http:/ers.usda.gov/publications/fanrr29/.

Nord, Mark, Margaret Andrews & Joshua Winicki, F. (2000). *Frequency and Duration of Food Insecurity and Hunger in U.S. Households*, paper presented at the Fourth International Conference on Dietary Assessment Methods, Tucson, AZ, Sept. 17-20, 2000.

Nord, Mark & Gary Bickel. (2002). *Measuring Children's Food Security in U.S. Households, 1995-99.* FANRR-25, USDA, Economic Research Service. Available at: http://webarchives.cdlib.org/sw15d8pg7m/http://ers.usda.gov/publications/fanrr25/.

Nord, Mark & Anne Marie Golla. (2009). *Does SNAP Decrease Food Insecurity? Untangling the Self-Selection Effect.* ERR-85, USDA, Economic Research Service. Available at: http://www.ers.usda.gov/ publications/err-economic-research-report/err85.aspx/.

Nord, Mark & Heather Hopwood. (2007). "Recent Advances Provide Improved Tools for Measuring Children's Food Security," *Journal of Nutrition*, *137*, 533-36.

Nord, M., Jemison, K. & Bickel, G. W. (1999). *Prevalence of Food Insecurity and Hunger by State, 1996-1998.* FANRR-2, USDA, Economic Research Service. Available at: http://webarchives.cdlib.org/ sw1m04028n/http://www.ers.usda.gov/publications/fanrr2/.

Nord, Mark & Kantor, Linda S. (2006). "Seasonal Variation in Food Insecurity Is Associated with Heating and Cooling Costs Among Low-Income Elderly Americans," *Journal of Nutrition*, *136*, 2939-44.

Ohls, James, Larry Radbill & Allen Schirm. (2001). *Household Food Security in the United States, 1995-1997: Technical Issues and Statistical Report.* Prepared by Mathematica Policy Research, Inc., for USDA, Food and Nutrition Service. Available at: http://www.fns.usda. gov/ora/menu/published/foodsecurity/foodsecuritytech.pdf/.

Oliveira, Victor. (2012). *The Food Assistance Landscape: FY 2011 Annual Report,* EIB-93, USDA, Economic Research Service. Available at: http://www.ers.usda.gov/publications/eib-economic-information-bulletin/eib93.aspx/.

Oliveira, Victor & Donald Rose. (1996). *Food Expenditure Estimates From the 1995 CPS Food Security Supplement: How Do They Compare With the Consumer Expenditure Survey?* Staff Report No. AGES9617, USDA, Economic Research Service.

Ratcliffe, Caroline & Signe-Mary McKernan. (2011). "How Much Does the Supplemental Nutrition Assistance Program Reduce Food Insecurity?" *American Journal of Agricultural Economics*, *93*(4), 1082-1098.

U.S. Department of Agriculture, Center for Nutrition Policy and Promotion. (2011). *Official USDA Food Plans: Cost of Food at Home at Four Levels, U.S. Average, December 2010*. Available at: http://www. cnpp.usda.gov/ publications/foodplans/2011/costoffooddec2011.pdf/.

U.S. Department of Agriculture, Center for Nutrition Policy and Promotion. (2007). *The Thrifty Food Plan, 2006*, CNPP-19. Available at: http://www. cnpp.usda.gov/publications/foodplans/miscpubs/tfp2006 report.pdf/.

Wilde, Parke & Mark Nord. (2005). "The Effect of Food Stamps on Food Security: A Panel Data Approach," *Review of Agricultural Economics*, *27*(3), 425-32.

Yen, Steven T., Margaret Andrews, Zhuo Chen & Eastwood, David B. (2008). "Food Stamp Program Participation and Food Insecurity: An Instrumental Variables Approach," *American Journal of Agricultural Economics*, *90*(1), 117-32.

End Notes

[1] See http://www.ers.usda.gov/top-ics/food-nutrition-assistance/food-security-in-the-us/history-background.aspx for the history of the food security measurement project and the development of the food security measures.

[2] Reweighting of the Supplement takes into consideration income and other information about households that completed the labor-force portion of the survey but not the Food Security Supplement. This corrects, to some extent, biases that could result from nonresponse to the Supplement by households that completed only the labor-force part of the survey.

[3] Standard errors of estimates, except for State-level estimates, were calculated using balanced repeated replication (BRR) methods based on household replicate weights computed for the CPS food security supplement by the U.S. Census Bureau (see http:// thedataweb.rm. census.gov/ftp/cps_ftp.html#cpsrepwgt). Standard errors of State-level estimates were calculated using jackknife replication methods with "month in sample" groups considered as separate, independent samples (see Nord et al., 1999). Beginning with the 2007 data, the jackknife methods have aggregated data from pairs of month-in-sample groups comprising largely the same households (i.e., month-in-sample group 1 in one year and month-in-sample group 5 in the following year) to account for the non-independence of these samples.

[4] The methods used to measure the extent and severity of food insecurity have been described in several places (Hamilton et al., 1997a, 1997b; Andrews et al., 1998; Bickel et al.,1998; Carlson et al., 1999; Bickel et al., 2000; Nord and Bickel, 2002). See also the assessment of the measurement methods by a panel of the Committee on National Statistics (National Research Council, 2006). Further details on the development of the measure are provided on the ERS website at: http://www.ers.usda.gov/topics/food-nutri-tion-assistance/food-security-in-the-us/history-background.aspx/.

[5] To reduce the burden on higher in-come respondents, households with incomes above 185 percent of the Federal poverty line that give no indication of food-access problems on either of two preliminary screening questions are deemed to be food secure and are not asked the questions in the food security assessment series. The preliminary screening questions are as follows:

- People do different things when they are running out of money for food in order to make their food or their food money go further. In the last 12 months, since December of last year, did you ever run short of money and try to make your food or your food money go further?
- Which of these statements best describes the food eaten in your household—enough of the kinds of food we want to eat, enough but not always the kinds of food we want to eat, sometimes not enough to eat, or often not enough to eat?

[6] Prior to 2006, households with low food security were described as "food insecure without hunger" and households with very low food security were described as "food insecure with hunger." Changes in these descriptions were made in 2006 at the recommendation of the Committee on National Statistics (National Research Council, 2006), in order to distinguish the physiological state of hunger from indicators of food availability. The criteria by which households were classified remained unchanged.

[7] Food security and insecurity, as measured for this report, are based on respondents' perceptions of whether the household was able to obtain enough food to meet their needs. The measure does not specifically address whether the household members' food intake was sufficient for active, healthy lives. Nonetheless, research based on other surveys has found food security, measured as in this report, to be associated with health, nutrition, and children's development in a manner that generally supports the conceptualized link with sufficiency for active, healthy lives (see, for example, Nord, 2009a; Nord and Hopwood, 2007; Nord and Kantor, 2006).

[8] Because of changes from year to year in screening procedures used to reduce respondent burden in the food security survey interviews, prevalence statistics calculated from the 1996 and 1997 data are not directly comparable with those for other years. The values presented in figure 3 for 1996 and 1997 are adjusted for the estimated effects of the differences in screening so as to be comparable with the statistics for other years. Screening procedures have remained essentially unchanged since 1998, and the procedures used in 1995 differed negligibly from those in 1998 and later years. See Andrews et al. (2000) and Ohls et al. (2001) for detailed information about questionnaire screening and adjustments for comparability. From 1995 to 2000, the prevalence rates reflected an overall decline in food insecurity but also a 2-year cyclical component that was associated with data collection schedules (Cohen et al., 2002). The CPS food security surveys over that period alternated between April in odd-numbered years and August or September in even-numbered years. The measured prevalence of food insecurity was higher in the August/September collections, suggesting a seasonal response effect. Since 2001, the survey has been conducted in early December, which avoids further problems of seasonality effects in interpreting annual changes. A smaller food security survey was also conducted in April 2001 to provide a baseline for assessing seasonal effects of data collection in December.

Comparison of food security statistics from the April 2001 survey with those from April 1999 and December 2001 suggest that seasonal effects in early December were similar to those in April (Nord et al., 2002).

[9] The Federal poverty line was $22,811 for a family of four in 2011.

[10] "Elderly" in this report refers to persons age 65 and older.

[11] About 40 percent of the difference in food insecurity between households with and without children results from a difference in the measures applied to the two types of households. Responses to questions about children as well as adults are considered in assessing the food security status of households with children, but for both types of households, a total of three

indications of food insecurity is required for classification as food insecure. Even with the child-referenced questions omitted from the scale, however, 17.2 percent of households with children would be classified as food insecure (that is, as having food insecurity among adults), compared with 12.2 percent for households without children. Comparisons of very low food security are not biased substantially by this measurement issue because a higher threshold is applied to households with children consistent with the larger number of questions taken into consideration.

[12] Some households with children headed by a single woman or a single man as classified for these analyses included other adults, who may have been parents, siblings, cohabiting partners, children, or other relatives of the reference person or unrelated roomers or boarders.

[13] Revised metropolitan statistical areas (MSAs) and principal cities within them were delineated by the Office of Management and Budget in 2003, based on revised standards developed by the U.S. Census Bureau in collaboration with other Federal agencies. Food security prevalence statistics by area of residence are comparable with those for 2004 and later years, but are not precisely comparable with those for earlier years. Principal cities include the incorporated areas of the largest city in each MSA and other cities in the MSA that meet specified criteria based on population size and commuting patterns.

[14] Households are classified as having food insecurity among children if they report two or more food-insecure conditions among children in response to questions 11-18 in box on page 3 (Nord, 2009a).

[15] Food spending is only an indirect indicator of food consumption. It understates food consumption in households that receive food from in-kind programs, such as the National School Lunch and School Breakfast Programs, the Special Supplemental Nutrition Program for Women, Infants, and Children (WIC), meal programs for children in child care and for the elderly, and private charitable organizations. (Purchases with SNAP benefits, however, are counted as food spending in the CPS food security survey.) Food spending also understates food consumption in households that acquire a substantial part of their food supply through gardening, hunting, or fishing, as well as in households that eat more meals at friends' or relatives' homes than they provide to friends or relatives. (Food spending overstates food consumption in households with the opposite characteristics.) Food spending also understates food consumption in geographical areas with relatively low food prices and over-states consumption in areas with high food prices.

[16] For spending in the first two categories of stores, respondents were also asked how much of the amount was for "nonfood items such as pet food, paper products, alcohol, detergents, or cleaning supplies." These amounts are not included in calculating spending for food.

[17] The cost of the Thrifty Food Plan is revised each month to account for inflation in food prices.

[18] The Thrifty Food Plan, in addition to its use as a research tool, is used as a basis for setting the maximum SNAP (food stamp) benefit amounts. However, in April 2009, the American Recovery and Reinvestment Act (ARRA) increased SNAP benefits above the Thrifty Food Plan-based levels. SNAP benefits have not been adjusted for inflation since then.

[19] Thrifty Food Plan costs are estimated separately for Alaska and Hawaii, using adjustment factors calculated from USDA's Thrifty Food Plan costs for those States for the second half of 2011.

[20] Households that were unable or unwilling to report food spending were less likely to be food insecure than those that did report food spending (9.6 percent compared with 15.3 percent). Food spending may, therefore, be slightly underestimated from these data.

[21] However, food spending does not rise proportionately with income increases, so high-income households actually spend a smaller proportion of their income on food than do low-income households.

[22] Information on Federal food and nutrition assistance programs, including participation rates and characteristics of participants, is available from the FNS website at http://www.fns.usda.gov/. Additional research findings on the operation and effectiveness of these

programs are available from the ERS website at http://www.ers.usda.gov/topics/food-nutrition-assistance.aspx/.

[23] The Food Stamp Program was renamed the Supplemental Nutrition Assistance Program (SNAP) in October 2008. Both names were mentioned in the survey question as well as the State's name for the program in States that used a different name.

[24] Some program participants reported incomes that were higher than the program eligibility criteria. They may have had incomes below the eligibility threshold during part of the year, or subfamilies within the household may have had incomes low enough to have been eligible.

[25] This "self-selection" effect is evident in the association between food security and food program participation that is observed in the food security survey. Participating households were less food secure than similar nonparticipating households. More complex analysis using methods to account for this self-targeting is required to assess the extent to which the programs improve food security (see Ratcliffe and McKernan, 2011; Nord and Golla, 2009; Yen et al., 2008; Wilde and Nord, 2005; Gundersen and Oliveira, 2001; Gundersen and Gruber, 2001; Nelson and Lurie, 1998).

[26] These statistics may be biased downward. It is known from comparisons between household survey data and administrative records that food program participation is underreported by household survey respondents, including those in the CPS (Meyer et al., 2009). This is probably true for food-insecure households as well, although the extent of underreporting by these households is not known. Statistics are based on the subsample of households with annual incomes below 185 percent of the poverty line. Not all of these households were eligible for certain programs. (For example, many households without pregnant women or children and with incomes above 130 percent of poverty would not have been eligible for any of the programs.)

[27] The statistics in table 8 were also calculated for households that were food insecure during the 30-day period prior to the survey. In principle, that analysis is preferable because food security status and use of programs are more certainly contemporaneous than when food insecurity is assessed over a 12-month period. However, the results differed only slightly from those in table 8 and are not presented separately.

In: Comparisons in Household Food Security
Editor: Samuel Owusu
ISBN: 978-1-62417-530-5

Chapter 2

ESTIMATING THE RANGE OF FOOD-INSECURE HOUSEHOLDS IN INDIA*

Sharad Tandon and Maurice R. Landes

ABSTRACT

This study provides a quantitative assessment of food security using a large household-level expenditure survey conducted by the Government of India during 2004/05. The analysis tests the impact of several key assumptions required to estimate actual calories consumed from the expenditure data. The authors found significant differences in the estimates of calories consumed and the number of food-insecure people under alternative plausible assumptions for computing the calorie content of nonprocessed foods, processed foods, and meals eaten outside the household. The measurement errors were largest in accounting for calories consumed by the highest and lowest income households. Overall, the difference between the highest and lowest estimate of the number of people consuming an average of less than 2,100 calories per day was equivalent to about 17 percent of India's population, or 173 million people in 2004/05. Given the significant measurement error in estimating calories consumed, it is important to consider not only consumption surveys, but also aggregate food availability studies and survey data on anthropometric measures that accompany undernourishment—such as growth stunting—in assessing food insecurity.

---

* This is an edited, reformatted and augmented version of the United States Department of Agriculture Publication, ERR-133, dated May 2012.

Keywords: Food insecurity, calorie consumption, nutrition, household survey, development, India

SUMMARY

What Is the Issue?

Policymakers and researchers are increasingly concerned with assessing the worldwide food-insecure population and the ways it may be changing. A common denominator in different approaches to assessing food insecurity is the measurement of calories consumed. The ability to measure food consumption is basic not only to gauging food insecurity, but also for targeting and evaluating policies aimed at alleviating it, such as the U.S. Government's Feed the Future initiative. This report examines problems in measuring calorie consumption and the food-insecure population, even when extensive household survey data are available, and finds that the extent of food insecurity varies markedly according to how it is measured. The analysis focuses on India, the country with the largest food-insecure population in the world, using a large household data set compiled by the Government of India for tracking household food security.

What Did the Study Find?

The authors found significant differences in estimates of the size of India's food-insecure population—comprising people who consumed less than 2,100 calories per day—across three major assessment methodologies: (1) The aggregate production and consumption approach used in the annual global food insecurity assessment by the U.S. Department of Agriculture (USDA); (2) the household expenditure survey approach; and (3) the survey of direct responses to questions on household food security status. Each of the methods, summarized briefly, describes a different but important aspect of food security:

- **Aggregate production and consumption approach**—Food production and trade statistics are used to determine a country's total food availability, which can then be used to estimate the number of food-insecure households through data on income distribution. Based on this approach, USDA's 2005 global assessment estimated India's food-insecure population at 217 million people.

- **Household expenditure survey approach**—Household responses to the Indian Government's National Sample Survey allow estimation of calories derived from food purchased or produced by households and also provide information on the characteristics of these households. Using these data and appropriate weighting factors to expand to the entire population, the authors found a baseline estimate of 508 million food-insecure people in India for 2004/05. However, the estimates ranged from 404 million to 577 million, depending on alternative plausible assumptions on the calorie content of foods and of meals consumed outside the home.
- **Survey of direct responses to questions on household food security**—This approach relies on survey questions that ask respondents about the adequacy of food consumption for household members. The authors found that estimates based on a specific question on household food security in India's National Sample Survey gave an estimate of 19 million food-insecure people in 2004/05, sharply lower than for the alternative methods. The authors note, however, that the Indian survey instrument differs significantly from the carefully designed, multi-question modules used in the United States and elsewhere.

India's household expenditure survey data may have the potential to provide the most accurate assessment of food insecurity, because they contain the most detail on household food availability. In analyzing the data, the authors found a large spread in estimates of the food-insecure population in India when they used alternative assumptions for estimating calories consumed. The highest and lowest counts of calorie intake resulted in an estimated difference of about 173 million food-insecure people in 2004/05. Specific measurement issues raised by the analysis are:

- Difficulty in determining calories in processed foods, an increasingly important component of diets across Indian households. Because of the wide range of nutritional content within various categories of processed foods, it is not possible to reliably discern calorie content from the survey data.
- Errors in estimating the calorie content of meals consumed outside the home and meals provided to nonhousehold members—growing trends

in Indian diets—for which calories cannot be precisely estimated from the survey data. Miscounting calories from these sources, as well as those from processed foods, can distort estimates of food insecurity rates across income groups and survey years.

- Potential errors associated with estimating consumption of processed foods and meals outside the home, which vary with household characteristics such as income and, therefore, are nonrandom sources of error. In this analysis, we find that the errors are largest when accounting for calories consumed by the highest and lowest income households.
- Conflicting sources of information on the calorie content of unprocessed foods. The calorie conversion factors used by the Government of India and FAO differ substantially for some foods.

How Was the Study Conducted?

Household data collected in the 2004/05 round of India's National Sample Survey, a survey of approximately 125,000 households conducted every 5 years, provide quantity and expenditure information for approximately 152 different food items for each household. Baseline estimates of calories purchased in the form of nonprocessed foods were calculated by multiplying quantities purchased by the average amount of calories per unit of quantity, using calorie conversion information from a source used by the Government of India. The baseline estimates of consumption of nonprocessed food calories were adjusted to account for household purchases of processed foods, calories included in the number of meals eaten outside the home, and meals given to nonhousehold members (such as to guests). The sensitivity of these baseline estimates of calorie consumption was then tested by perturbing the baseline assumptions, including (1) using an alternate source of information on the calorie content of foods from the Food and Agriculture Organization of the United Nations, (2) using alternate assumptions on the cost of calories from processed foods, and (3) accounting for the error involved in estimating the calories included in meals consumed outside the home or given to nonhousehold members.

INTRODUCTION

Despite India's rapid income growth over the past two decades, current estimates of the number of food-insecure people derived from aggregate production, trade, and income distribution data suggest that the country accounts for nearly 40 percent of the world's food-insecure population (e.g., Shapouri et al., 2011). This estimate is corroborated by a high prevalence of anthropometric indicators that accompany undernourishment, such as being underweight (e.g., Deaton and Dreze, 2008). Even though there is overwhelming evidence that a large share of the Indian population does not have adequate access to food, quantifying the extent of the problem and its change over time remains problematic not only in India, but also in other developed and developing countries (Nord et al., 2008; Barrett, 2010).

Three alternative approaches are commonly used to assess national food security status:

1. Computing average calorie consumption from aggregate food supply and use balances.
2. Using household survey data on food expenditures to estimate household-level food consumption and food security status.
3. Using survey data on household members' assessments of the adequacy of food availability to determine their food security status.

Policymakers often focus on estimates of calories consumed in the first two approaches. Each of these methods describes a different but important aspect of food security. Aggregate production and trade statistics better describe food availability, which can then be used to derive estimates of the number of food-insecure households by using information on the distribution of incomes (Shapouri et al., 2011). Household-level consumption survey data, by comparison, generally do not provide information on the sources of consumption (e.g., domestic versus imported food), but allow policymakers to identify which individual households are food insecure, along with the location and other characteristics of those households. Household survey data also generally allow for a more complete and detailed accounting of food items purchased, resulting in somewhat more precise estimates of the nutritional content of food consumed by each individual household (Deaton and Subramanian, 1996).

The third approach for assessing food consumption and food insecurity involves using a carefully designed and administered survey module to ask respondents directly about the adequacy of food consumption for household members (National Research Council, 2006). This is the approach used by USDA for assessing food security in the United States. It has also been used in local or national surveys in other countries, including Argentina, Bangladesh, Brazil, India, and Uganda, and is being developed for use by the U.S. Agency for International Development to track food security conditions in developing countries (Nord et al., 2002; Deitchler et al., 2011). This method bypasses the data problems of estimating the nutritional content of foods consumed, but it does not provide information on food sources or diet composition and may yield different results in different cultural settings (Saxena, 2011).

The Government of India regularly generates estimates of food security status for policymakers, using survey-based estimates of caloric intake. This report, taking advantage of these data, focuses on the calorie consumption of individual Indian households using a household-level survey. There have been a number of estimates of calorie purchases in India using similar data (e.g., Deaton and Subramanian, 1996; Deaton and Dreze, 2008; National Sample Survey Organization, 2007; Tandon and Landes, 2011). While estimates of calorie purchases are based on the sum of calories in nonprocessed and processed food items, adjusted for calories contained in meals given to nonhousehold members and consumed in meals outside the home, these estimates require a number of important assumptions that, according to our findings, can significantly affect the precision of the results.

Next, we briefly describe the survey used to estimate calorie consumption throughout this report. We then explain the methods we used to estimate calorie consumption from the survey and calories consumed by each household. Subsequently, we examine the sensitivity of the baseline estimates to the use of alternative assumptions for three key components of the estimates: (1) the calorie content of nonprocessed foods, (2) the calorie content of processed foods, and (3) the calories in meals consumed outside the home and provided to nonhousehold members. Finally, we assess the overall findings and provide comparisons with food insecurity estimates from alternative approaches.

DATA

Household-level expenditure data are collected annually by India's National Sample Survey Organization (NSSO). The NSSO conducts annual surveys on a range of topics and conducts a more detailed survey of both consumer expenditures and employment every 5 years. These surveys are the official source of household data used by the Indian Government to monitor and study poverty and food security in the country. For the consumer expenditure surveys, a sample of approximately 125,000 households is surveyed in each round, and the data collected include quantity and value of purchases of approximately 152 separate food items, along with the sources of each item (e.g., home-produced, purchased, etc.). The survey also reports the number of meals consumed outside the household and provided to nonhousehold members. In addition to household consumption, the survey reports a range of household and individual characteristics, including the number of household members, the household location, and the education and age of household members. The 61st round of the NSSO Consumer Expenditure Survey used in this study reports detailed expenditure data for 124,536 households collected during 2004-05.[1]

The NSSO Consumer Expenditure Survey is not a random sample of Indian households Rather, the sample is geographically dispersed and stratified into rural and urban portions and further stratified based on measures of income. Portions of rural villages and urban towns are randomly selected to sample households and, within these regions, households of particular income groups and sectors (e.g., rural or urban) are randomly sampled. Thus, the sample becomes representative of the entire country only when the data for the households in the survey are properly weighted based on their prevalence in the entire country.[2]

Estimation of Calorie Consumption from Household Expenditures

We first estimate household calorie consumption using an approach developed by Deaton and Subramanian (1996) for estimating calorie consumption from household expenditure data. It is important to note that the calorie consumption estimates are derived from the NSSO data that include both expenditures on purchases outside the household and food produced at home. The estimates do not, however, account for food losses that may occur at the household level. The estimation strategy involves three steps:

- Estimation of calories consumed from nonprocessed foods, such as particular types of rice, fruits, or vegetables. In order to estimate calories consumed from each source, we follow Deaton and Subramanian and use the estimated calories contained in quantities of each food, based on a nutritional study of a broad array of Indian foods conducted by the Government of India (Gopalan et al., 1989).
- Estimation of calories consumed from processed foods such as salted snacks, cakes, and pastries. Because of the wide variety of items in this category, it is difficult to estimate the actual number of calories in each item.[3] Given this variety, Deaton and Subramanian based their estimates on the value of expenditures on processed foods and used the simplifying assumption that calories from processed foods are purchased at a premium over nonprocessed foods. Specifically, they calculated the average calories purchased per rupee spent for each household and assumed that households received 50 percent fewer calories per rupee spent on processed foods than they did from nonprocessed foods.
- Estimation of calories consumed in meals outside the household and in meals given to nonhousehold members.[4] The NSSO survey provides information on the number of such meals for each household, but it is necessary to impute the number of calories associated with these meals. Deaton and Subramanian accomplished this by econometrically estimating the relationship between the total nonprocessed and processed calories purchased by each household, the numbers of home meals, meals given, and meals outside the home. Those results are then used to derive estimates of the calories associated with each type of meal. Under restrictive assumptions, this methodology provides an estimate of calories contained in each type of meal.

Once the baseline estimates of total calorie consumption are calculated, calorie intake totals for individuals are computed to permit comparisons with individual consumption benchmarks. This computation must account for both size and composition of households. To achieve this, we use the age and gender of children and adults to adjust total household size to "young adult equivalents," for which the benchmark consumption level—used in the 2005 USDA Food Security Assessment to define a food-insecure person—is 2,100 calories per day.[5,6]

BASELINE ESTIMATES OF CALORIE CONSUMPTION

The first step in generating the baseline estimates is computation of calories derived from the nonprocessed foods itemized in the survey household data, using the conversion factor found in Gopalan et al. (1989). The results for calories from nonprocessed foods are shown in table 1.

Table 1 also summarizes information on expenditures on processed foods, the number of meals consumed outside the home, and meals given to nonhousehold members, all of which will be accounted for to derive an estimate of total calories purchased or home-produced. In the case of processed foods, for which only value information is available in the NSSO data, household members spent an average of 3.79 rupees per day. This is equivalent to about 9 percent of daily food expenditures, although it is substantially larger for some households. Meals consumed outside the home—including meals paid for, received from employers, consumed at school, and a residual category of "other meals received" (including meals at weddings, birthdays, and other ceremonies)—do not appear to constitute a large category, on average, but some households do consume a large number of these types of meals. The number of meals provided to nonhousehold members also appears small for the average household, but again, the values become large for some households.

The results for nonprocessed calories purchased can also be broken out both by food category and the income groups identified in the NSSO survey (table 2). Column 1 in table 2 shows summary statistics for households with income in the top 10th percentile, and column 2 presents summary statistics for all other households, labeled as "middle income and poor," with results pooled across both urban and rural sectors.[7] The differences in column 3 indicate that poorer households purchase an average of 315 fewer calories than their more affluent counterparts and also purchase significantly less expensive calories, as indicated by the 51 fewer calories received per rupee spent.

Regardless of income level, the majority of nonprocessed calories purchased are derived from grain products. Despite the difference in total nonprocessed calories purchased between poorer and richer households, richer households are shown to purchase nearly identical amounts of grains—statistically, the hypothesis that grain purchases are equal between the two groups cannot be rejected at the 10-percent significance level. However, more affluent households purchased significantly more of every other food category, with the exception of calories from alcohol. This difference in purchases of foods that are more expensive partially accounts for the substantial difference between rich and poor households in rupees spent per calorie purchased.[8,9]

Calories from Processed Foods

Processed foods comprise a broad array of products of varying nutritional content for which the NSSO survey provides only expenditure—but not quantity—data. Processed food categories in the survey include items such as "salted refreshments," "cake/pastry," and residual categories such as "other milk products" and "other processed food," for which it is difficult to assign calorie values.[10] For our analysis, rather than attaching a calorie value to a quantity purchased, it is necessary to attach a calorie value to household expenditures on these processed foods. The baseline estimates follow Deation and Subramanian (1996) by assuming that the cost of calories from processed foods is twice the cost of calories from nonprocessed foods.

Following this approach, the cost of calories from nonprocessed foods is computed for each household and then doubled to convert each household's expenditures on processed foods to a calorie equivalent.

Table 1. Summary of data used in constructing estimates of household calorie consumption

Variable	Average	Std. deviation	Minimum	Maximum	Observations
Nonprocessed calories purchased per household member	2,107	845.4	0	68,993	122,888
Rupees spent by entire household on processed food items per day	3.79	4.46	0	181.3	122,888
Meals eaten outside by the entire household per day	.529	.984	0	21.3	122,888
Meals given by household to nonmembers per day	.021	.366	0	32.2	122,888

Source: Estimates calculated by authors using the 61st round of the National Sample Survey.

Table 2. Summary statistics of nonprocessed calories purchased per day by income group

Variable	Affluent households	Middle-income and poor households	Difference (column 1– column 2)
	1	2	3
Calories purchased per household member per day	2,356 (6.29)	2,041 (2.51)	315.1*** (9.52)
Calories per rupee spent	228.8 (.841)	279.8 (.480)	-51.0*** (1.51)
Source of nonprocessed calories			
Calories from grains	1,420 (3.77)	1,424 (1.83)	-4.96 (6.80)
Calories from pulses	116.6 (.597)	87.6 (.205)	29.0*** (1.05)
Calories from butter/oils/sugar	333.8 (1.95)	239.9 (.566)	93.9*** (2.87)
Calories from meat	25.5 (.308)	18.3 (.157)	7.18*** (.653)
Calories from vegetables	104.4 (.504)	89.3 (.252)	15.1*** (.775)
Calories from fruits and nuts	79.1 (1.09)	39.9 (.352)	39.2*** (1.69)
Calories from eggs	7.64 (.091)	4.97 (.03)	2.67*** (.172)
Calories from dairy	267.3 (1.61)	133.7 (.638)	133.5*** (3.61)
Calories from alcohol	2.28 (.186)	2.88 (.108)	-.594*** (.221)
Observations	25,730	97,158	

Standard errors clustered by district are reported in parentheses.
*** Denotes significance at the 1% level; **Denotes significance at the 5% level;
* Denotes significance at the 10% level.
Source: Estimates calculated using the 61st round of the National Sample Survey.

Although each household has different values for both calories purchased per rupee and rupees spent on processed foods, the sample averages are 3.79 rupees per household spent on processed foods per day and a cost of 269 calories of nonprocessed foods per rupee. Applying the assumption that processed calories cost twice as much as nonprocessed calories, the average household purchases approximately 508 processed calories per day, or 114 calories per household member per day. It is important to note, however, that there does not appear to be a strong basis for assuming that the average processed calorie costs twice as much as the average nonprocessed calorie. It is possible that processed foods vary widely in caloric content and cost and that the composition of purchases of processed foods varies across households based on such factors as occupation (agricultural versus nonagricultural), location (rural versus urban), and income level.

Calories from Whole Meals Received or Given

The NSSO data provide information on the number of meals consumed outside the household and meals provided to nonhousehold members, so it is necessary to impute estimates of the calories associated with those meals to make the required additions and subtractions to household calorie totals. Following the methodology introduced by Deation and Subramanian (1996), the baseline of calories contained in various types of meals is estimated with following specification:

$$Calories_{ir} = \tau_r + \beta_1 HomeMeals_{ir} + \beta_2 MealsGiven_{ir} + \beta_3 MealsReceived_{ir} + ControlVars_{ir} + \varepsilon_i$$

$Calories_{ir}$ denotes the total amount of calories purchased over the past 30 days, both processed and nonprocessed, by household i in district r; τ_r denotes a district fixed effect to help absorb unobserved characteristics shared by all households within a district; *HomeMeals* denotes the total number of meals eaten by household members at home over the past 30 days; *MealsGiven* denotes the total number of meals the household served to nonhousehold members over the past 30 days; *MealsReceived* denotes the total number of meals consumed by household members outside of the home over the past 30 days, which is the sum of meals received from school, employers, other sources, and on payment; and *ControlVars* denotes a number of household characteristics that are used as control variables to help absorb unobserved variation in calories purchased.[11,12]

Although this approach requires strong assumptions, the coefficients from this specification can be used to adjust calories purchased to account for each different type of meal. The coefficient β_2 provides an estimate of the number of the total calories purchased that were consumed by nonhousehold members for each meal, and β_3 provides an estimate of the number of calories consumed by household members in meals they ate outside the household.

Estimates of the specification are presented in table 3. Column 1 provides the estimates for a sparse specification, excluding the district fixed effects and control variables. Relative to all other estimated coefficients, the estimated coefficient for meals consumed at home has a narrow 95-percent confidence interval of approximately 11.0 calories; the coefficient for *Meals Given* is also statistically significant, but has a much larger 95-percent confidence interval.[13] On the other hand, *Meals Received* is not statistically significant and has an implausibly small magnitude for calories consumed in a meal.

To investigate whether this is true for all types of meals received outside the household, column 2 shows the variable broken down into the number of meals received at school, from employers, and from other sources and meals received for payment. The coefficients on meals eaten at home and meals given to nonhousehold members are essentially unchanged. On the other hand, coefficients on meals from school, received for payment, and received from other sources are statistically significant. However, the signs on the coefficients differ. Meals from school have the expected sign and suggest that each meal consumed at school could contain approximately 400 calories, which is the resulting decrease in calorie purchases for the household per school meal.

The signs on the coefficients of meals received for payment and meals from other sources are the opposite of what would be expected when holding all other factors fixed, and the coefficient is likely picking up the effects of some omitted variables on the total calories purchased. For example, given that these meals include meals at ceremonies such as weddings, it could be that primarily richer households receive these meals at ceremonies, and thus the coefficient is partially capturing the effects of income on calories purchased. Or perhaps receiving a meal from another household requires the household to provide a meal to that guest in return on some other occasion.

The results in columns 3 and 4 try to account for some unobserved heterogeneity by estimating specifications, including fixed effects and control variables. The coefficients on types of meals received vary significantly in magnitude and, in one instance, even change signs. Given these changes, it is difficult to identify the number of calories contained in meals received outside the home. However, the estimates of the amount of calories associated with meals given by households seem to be stable in magnitude and statistically significant across specifications.

Thus, when adjusting the calories purchased by individual households in response to meals eaten outside the home and meals given to nonhousehold members, we use the estimated coefficient on meals given in column 4 (664 calories per meal). However, it should be noted that, in that estimation, one cannot reject the hypothesis that true calorie content of a meal is anything between 221 and 1,107 calories at the 5-percent significance level.[14,15]

Table 3. Calories purchased per meal consumed

Type of meal	**Dependent variable: Total household calories purchased**			
	1	**2**	**3**	**4**
Meals at home	694.6*** (7.84)	701.2*** (8.03)	696.2*** (6.98)	370.1*** (7.93)
Meals given	827.0*** (269.9)	818.3*** (268.8)	818.8*** (265.1)	664.7*** (225.6)
Meals received	49.2 (34.3)			
Meals received from school		-401.2*** (65.1)	-327.9*** (37.2)	-158.0*** (35.9)
Meals received from employer		-135.5 (96.7)	-188.4*** (50.1)	-178.8*** (52.7)
Meals received for payment		207.0** (99.8	378.8*** (52.8)	-111.4* (67.6)
Meals received from other sources		349.9*** (42.1)	365.9*** (27.3)	49.8** (22.4)
District fixed-effects	N	N	Y	Y
Control variables	N	N	N	Y
Observations	122,888	122,888	122,888	122,888

Notes: * Denotes significance at the 10% level; ** Denotes significance at the 5% level;
*** Denotes significance at the 1% level. Standard errors are reported in parentheses.
Source: Estimates calculated using the 61st round of the National Sample Survey.

Total Calories Purchased per Household Member

Based on the estimation strategies described above, table 4 presents summary statistics of household calorie consumption by source. The vast majority of calories are obtained from nonprocessed foods. Other sources contribute to overall calorie consumption but constitute a much smaller share. However, some households consume a large number of calories from these other sources, suggesting that calories from processed foods, meals eaten outside the household, and given in meals to nonhousehold members are important to any estimate of food insecurity.

Accounting for calories consumed from all sources in table 4, table 5 provides estimates of total calorie consumption (column 1) and the share of all households in the sample consuming below 2,100 calories per capita per day (column 2). Columns 3 and 4 estimate the same parameters for the entire Indian population using probability weights. The estimate of average daily calories consumed per household member in the sample is approximately 2,303 (column 1). Of this total, approximately 114 calories on average were obtained from processed foods and approximately 85 calories from meals eaten outside the household.

According to these baseline estimates, approximately 47.2 percent of the households in the sample consumed less than 2,100 calories per person day, the consumption level used to define a food-insecure person in the USDA's Food Security Assessment for 2005 (Meade et al., 2006).[16] If we extrapolate this result to all of India, the average person consumes approximately 2,225 calories (column 3), and approximately 49.5 percent of households consume fewer than 2,100 calories per person per day. This share of households translates to approximately 508 million people according to the 2001 Indian Census.[17] The differing results for the sample households (columns 1 and 2) and the country totals (columns 3 and 4) reflect the use of weights in the national totals to compensate for the NSSO sampling procedure, in which richer households are oversampled relative to their prevalence in the entire population.[18]

Figure 1 provides the histogram of average daily per capita calorie purchases and describes the entire distribution of the estimated calorie purchases. The distribution is centered at the mean calorie purchases of 2,303 calories per day presented in table 4 and is skewed slightly to the right-hand side, implying that there are more people consuming well above the average calories than well below the average.

Table 4. Summary statistics of calories consumed per day by source

Variable	Average	Std. deviation	Minimum	Maximum	Observations
Nonprocessed calories per household member	2,107	845	0	68,993	122,888
Processed calories per household member	114	302	0	76,730	122,888
Calories consumed in meals outside the household per household member	85	198	0	3,988	122,888
Calories given to nonhousehold members per household member	3.6	80	0	16,817	122,888

Source: Estimates calculated by authors using the 61st round of the National Sample Survey. Each cell reports unweighted summary statistics from the entire sample of 122,888 households.

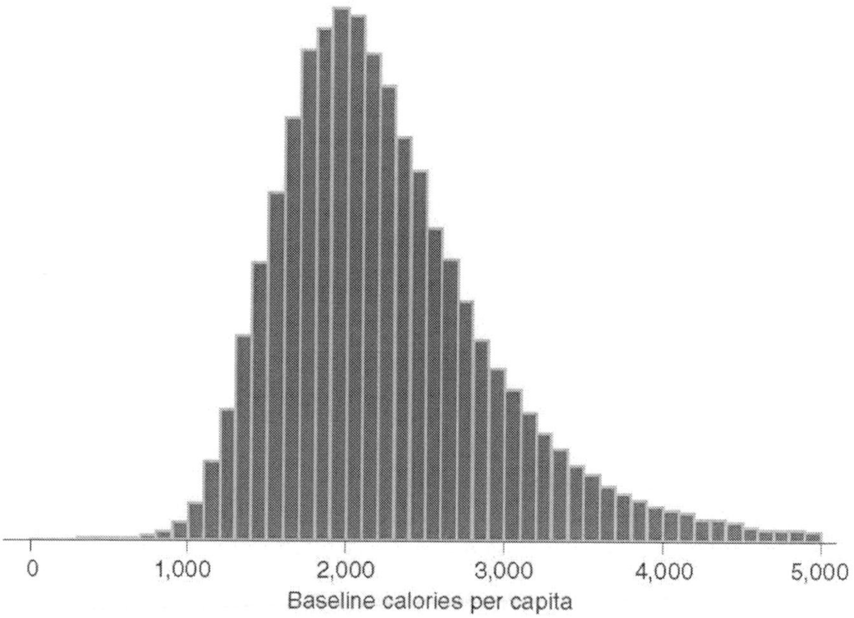

Source: Estimates calculated using the 61st round of the National Sample Survey.

Figure 1. Distribution of daily per capita calories purchased.

Table 5. Baseline estimate of average calories consumed

	All sample households		Total country estimates	
	1	2	3	4
Variable	Avg. per capita cal./day	Share of sample that is food insecure	Avg. per capita cal./day	Share of food-insecure households
Total calories consumed	2,303 (2.59)	.472 (.001)	2,225 (4.27)	.495 (.003)
Number of households	122,888			

Note: Standard errors are reported in parentheses.

Source: Estimates calculated using the 61st round of the National Sample Survey. Population estimates are constructed using National Sample Survey Organization (NSSO) multipliers to calculate population weights.

RESULTS USING ALTERNATE ASSUMPTIONS

The next step in the analysis is to evaluate the impacts on the individual baseline caloric intake estimates of using alternative plausible assumptions for deriving those estimates. Each step in computing the baseline calorie estimates required assumptions or methods that are potential sources of error in the results. In this section, the baseline assumptions are replaced with other plausible assumptions, one by one, to examine the impact on the estimates of calories purchased and the food-insecure population. The assumptions and methods used in each of the three steps of the baseline calculations are examined by:

- Using alternative sources of data on the calorie content of nonprocessed foods.
- Using alternative assumptions on the costs of calories from processed foods.
- Accounting for the large confidence intervals around the econometric estimates of calories associated with meals eaten outside the household and given to nonhousehold members.

Adjustment 1. The effect of alternative sources of data on the calorie content of nonprocessed food.

Estimating calorie consumption from household survey data requires the use of conversion factors for converting food quantity data into calorie equivalents. The baseline estimates of nonprocessed calories consumed per household member are based on the nutritional information for Indian foods contained in the *Nutritive Value of Indian Food* (NVIF) (Gopalan et al., 1989). To evaluate the sensitivity of these results to the choice of conversion factors, calorie consumption was also calculated using the conversion factors employed by the Food and Agriculture Organization of the United Nations (FAO) for estimating aggregate calorie availability for South Asian countries.[19]

The two sources differ in the types of food categories for which nutritional information is provided. NVIF provides nutritional information for more disaggregated food items than the FAO source, which applies average calorie information for a number of different food items.[20] Neither source will yield unambiguously higher estimates of total calories purchased. NVIF lists higher calorie information for 59 food items, FAO provides higher calorie information for 40 food items, and 53 items have the same calorie value. The net effect of using one source over the other depends on which food items are more common in a household's food purchases.

Although NVIF is somewhat more detailed, the comparison with FAO sources is informative for two reasons. First, FAO estimates the number of food-insecure people in India (e.g., FAO, 2011), and it is important to assess the possible reasons why estimates might differ between sources. Second, the comparison can demonstrate whether averaging across food items might obscure the numbers of food-insecure households in a country.

To examine the sensitivity of the results to the alternative sets of conversion factors, the total nonprocessed calories consumed are computed using the two sets of calorie information. We then estimate the following empirical specification:

$$NonprocessedCalories_{ir} = \tau_r + \beta FAONonProcessedCalories_{ir} + \varepsilon Controls_{ir} + \varepsilon_i$$

where $NonprocessedCalories_{ir}$ denotes the total number of nonprocessed calories consumed in household i and district r when calories are calculated using NVIF conversions, and *FAONonprocessedCalories* denotes the same measure with calories calculated using the nutritional information provided by FAO. The symbol τ denotes fixed effects, introduced to account for time-invariant differences between districts, and *Controls* denotes a number of household variables that are used to control for the effects of observable household characteristics on calories consumed.[21] With this approach, the resulting estimates of the coefficient β will describe how nonprocessed calories consumed differ between the two sources. Estimates of calorie consumption using the nutritional information in NVIF will be higher if the value of β is greater than 1 (β>1) and, conversely, will be lower if β is less than 1(β<1).

The results are shown in table 6. Column 1 presents estimates for a sparse specification of the model with no fixed effects or control variables; column 2 presents estimates when the district fixed effects are added; and column 3 presents estimates when both district fixed effects and household control variables are included. Because the estimates of β are greater than 1 and the p-value from the hypothesis test of β=1 is essentially 0 in all specifications, the results indicate that using the FAO conversion factors yields lower estimates of nonprocessed calories consumed than those derived using the NVIF factors. The results in column 3, the most complete estimation model, suggest that increasing the FAO calories consumed by one calorie is associated with an increase in NVIF calories by 1.031, a difference of 3.1 percent.

Table 7 shows the estimates of average total calorie consumption and the share of households purchasing less than 2,100 calories per day using both the NVIF (column 1) and FAO (column 2) conversion factors. Column 3 indicates that per capita calorie consumption is approximately 63 calories lower when using FAO conversion factors to calculate nonprocessed calories consumed rather than the NVIF conversion factors. As a consequence, the estimated share of food-insecure households is about 3.6 percent higher when using the FAO conversion factors. These estimates from the sample are mirrored by estimates for the entire population in columns 4 through 6. For the population as a whole, the FAO calorie information increases the estimate of food-insecure households by 5.1 percent, which translates into approximately 53 million people according to the 2001 Indian Census (column 6).[22]

Adjustment 2. Using alternate assumptions on the cost of calories from processed foods

Following the estimation strategy used by Deaton and Subramanian (1996), the baseline estimates of household calorie consumption assumed that calories from processed food in each household are twice as expensive as the calories from nonprocessed food. This assumption has both desirable and undesirable properties. In particular, the assumption implies that poorer households also purchase cheaper processed foods than richer households, which is a plausible characterization of actual consumption behavior. However, there does not appear to be a strong rationale supporting the assumption that processed calories should be exactly twice as expensive as nonprocessed calories, making it a source of uncertainty in the resulting estimates of household calorie purchases.

The importance of the assumed calorie content of processed foods is assessed by analyzing alternate scenarios. First, we assume that calories purchased in processed foods are the same price as those purchased in nonprocessed foods, and then we assume that processed food calories are four times more expensive. Although these assumptions are simplistic, many varieties of processed foods purchased by the average household are both more and less expensive than these bounds, and these likely represent only a fraction of the processed food items available to Indian consumers. The bounds chosen are intended to illustrate a range of possibilities rather than a best estimate of calories consumed from processed foods. Until household data can identify what processed foods households are actually purchasing and their caloric content, there will be a large confidence interval on any estimate of calories purchased from processed foods.

Table 6. Correlation between calories consumed using NVIF nutritional information and FAO nutritional information

	Dependent variable: Nonprocessed calories purchased using NVIF		
	1	2	3
Nonprocessed FAO calories	1.034*** (.003)	1.033*** (.003)	1.031*** (.006)
Fixed effects	N	Y	Y
Control variables	N	N	Y
P-value of test coefficient on FAOCal=1	.0000	.0000	.0000
Observations	122,888	122,888	122,888

NVIF=*Nutritive Value of Indian Foods* (Gopalan et al., 1989); FAO=Foreign Agriculture Organization.

*** Denotes significance at the 1% level; **Denotes significance at the 5% level; * Denotes significance at the 10% level. Standard errors are reported in parentheses.

Source: Estimates calculated using the 61st round of the National Sample Survey.

Table 7. Comparison of calorie consumption using alternate sources of nutritional information

	All sample households			Total country estimates		
	1	2	3	4	5	6
	Calories purchased using NVIF	Calories purchased using FAO	Difference	Calories purchased using NVIF	Calories purchased using FAO	Difference
Average calorie consumption	2,303 (2.59)	2,240 (2.53)	63.3*** (.375)	2,225 (4.27)	2,147 (4.06)	77.7*** (.736)
Share of food-insecure households	.472 (.001)	.508 (.001)	-.036*** (.001)	.495 (.003)	.546 (.003)	-.051*** (.001)

NVIF=Nutritive Value of Indian Foods (Gopalan et al., 1989); FAO=Foreign Agriculture Organization.

Notes:

1. Standard errors clustered at the district level are presented in parentheses.
2. *** Denotes significant at the 1% level; **significant at the 5% level; * significant at the 10% level.
3. Columns 1 and 4 calculate the estimate of calories purchased using NVIF nutritional totals and the share of households that purchased fewer than 2,100 calories. Columns 2 and 5 calculate the same totals using FAO nutritional tools. Columns 3 and 6 present the difference between the two, subtracting columns 2 and 5 from columns 1 and 4, respectively.

Source: Estimates calculated using 122,888 households from the 61st round of the National Sample Survey. Population estimates are constructed using National Sample Survey multipliers to calculate population weights.

The results show that the alternate assumptions on the cost of processed food calories have a significant impact on the estimates of per capita calorie consumption and the food-insecure population (table 8). The difference between assuming that processed calories are four times as expensive as those in nonprocessed foods and assuming they are the same price as for nonprocessed foods is an average of 171 calories per capita (column 3). That translates into a 9.5-percent difference in the share of the sample households with per capita daily purchases less than 2,100 calories (column 3). If we extrapolate to the entire population in column 6, the estimated difference in the number of people purchasing less than 2,100 calories per day is over 98 million people based on the 2001 Indian Census.

Table 8. Comparison of calorie consumption using alternate assumptions for the cost of processed foods

	All sample households			Total country estimates		
	1	2	3	4	5	6
Variable	Consumption assuming processed and nonprocessed foods cost the same	Consumption assuming processed foods cost four times as much as nonprocessed foods	Difference	Consumption assuming processed and nonprocessed foods cost the same	Consumption assuming processed foods cost four times as much as nonprocessed foods	Difference
Average calorie consumption	2,417 (3.05)	2,246 (2.44)	171.2*** (1.29)	2,325 (4.85)	2,174 (4.12)	150.5*** (2.09)
Share of food-insecure households	.410 (.001)	.504 (.001)	-.095*** (.001)	.433 (.003)	.528 (.003)	-.095*** (.002)

Notes:

1. Standard errors are presented in parentheses. 2. *** Denotes significance at the 1% level; ** denotes significance at the 5% level; * denotes significance at the 10% level. 3. Columns 1 and 4 estimate the average calories purchased and the share of households that purchased less than 2,100 calories using an estimate of calorie consumption that assumes processed foods are the same price as nonprocessed foods. Columns 2 and 5 calculate the same totals using an estimate of calorie purchases that assumes processed foods are four times as expensive as nonprocessed foods. Columns 3 and 6 present the difference between the two, subtracting columns 2 and 5 from columns 1 and 4, respectively.

Source: Estimates calculated using 122,888 households from the 61st round of the National Sample Survey. Population estimates are constructed using NSS multipliers to calculate population weights.

Because of the potential errors introduced by oversimplified assumptions about the nutritional content of processed foods, it would be valuable for future studies to combine the NSSO consumption data with other data sources that better describe the processed foods most consumed by various categories of Indian households, as well as the calorie content of those foods. For example, Euromonitor International reports total expenditures on all packaged foods sold by grocery retailers, as well as totals by subcategories, such as "salted snacks," that correspond to some of the NSSO codes of processed foods.[23] The same source also reports the shares of each of these subcategories captured by particular brands, which may permit collection of more precise information on prices and nutritional content.

Adjustment 3. Accounting for the large confidence limits around estimates of calories in meals eaten outside the household or given to nonhousehold members

As discussed earlier, adjusting household calories purchased for meals consumed outside the household and meals given to nonhousehold members requires estimation of the calories consumed per meal. The NSSO expenditure surveys do not provide information needed to directly compute calories consumed in such meals, and the econometric procedures used by others (e.g., Deaton and Subramanian, 1996) yield imprecise estimates and introduce potential errors.

To evaluate the importance of this problem, we assess the impacts of a range of alternate assumptions about the calories in meals consumed outside the household or given to nonhousehold members by accounting for the uncertainty in the regression techniques used to estimate the calories contained in both types of meals. The regression estimate of the calories included in both types of meal is approximately 664 calories (table 3), but if we require a 5-percent significance level, we cannot rule out a true calorie value anywhere inside the 95-percent confidence limits of this estimate, or between 221 and 1,107 calories per meal.

Table 9 provides estimates of average calories consumed and the share of food-insecure individuals for the sample and population, when we use the upper and lower bound confidence limits around the econometric estimate of calories per meal. These results indicate that the difference between using the upper and lower bound limits has an impact on the share of food-insecure households similar to that from using alternate sources of caloric conversion information. The difference in the share of the population that is food insecure between estimates of calorie purchases using the upper and lower bound confidence limit is approximately 7.4 percent (column 6), which translates into nearly 76 million people according to the 2001 Indian Census.

Table 9. Comparison of calorie purchases using alternate assumptions for meals eaten outside the household

	All sample households			Total country estimates		
	1	2	3	4	5	6
Variable	**Consumption using upper bound limit for meal estimate**	**Consumption using lower bound limit for meal estimate**	**Difference**	**Consumption using upper bound limit for meal estimate**	**Consumption using lower bound limit for meal estimate**	**Difference**
Average calorie consumption	2,357 (2.63)	2,248 (2.60)	108.9 *** (.813)	2,279 (4.28)	2,170 (4.34)	108.8*** (1.23)
Share of food-insecure households	.437 (.001)	.504 (.001)	-.067*** (.001)	.458 (.003)	.532 (.003)	-.074*** (.001)

Notes:

1. Standard errors are presented in parentheses.
2. *** Denotes significance at the 1% level; ** denotes significance at the 5% level; * denotes significance at the 10% level.
3. Columns 1 and 4 calculate the estimate of calories purchased and the share of households that purchased less than 2,100 calories using the upper bound of the estimate of calories contained in meals eaten outside the household. Columns 2 and 5 calculate the same totals using the lower bound of the estimate of calories contained in meals eaten outside the household. Columns 3 and 6 present the difference between the two, subtracting columns 2 and 5 from columns 1 and 4, respectively.

Source: Estimates calculated using 122,888 households from the 61st round of the National Sample Survey. Population estimates are constructed using NSS multipliers to calculate population weights.

ADDING IT ALL UP: IMPLICATIONS FOR COUNTING THE FOOD-INSECURE POPULATION

The preceding sections provided baseline estimates of average caloric intake levels and the size of the food-insecure population, along with plausible alternative estimates for the key components of those calorie totals—nonprocessed foods, processed foods, and meals eaten or given outside the home—that suggest scope for substantial error in the baseline estimates. Each of the alternative scenarios yielded outcomes that are significantly different from each other. Based on the scenario findings, it is possible to construct "High" and "Low" estimates of India's food security status consistent with the NSSO data by combining the scenarios that yield the lowest and highest estimates of individual calories consumed. The following specific assumptions are made in constructing the High and Low estimates of calories consumed:

- *High-calorie estimate:* Uses the calorie conversion factors provided in the list used by the Indian Government, the *Nutritive Value of Indian Foods* (NVIF), to estimate calories in nonprocessed food purchases; assumes that processed calories are the same price as nonprocessed calories; and uses the upper bound of the estimate of calories consumed in meals outside the home and given to nonhousehold members.
- *Low-calorie estimate:* Uses the calorie conversion factors used by the Food and Agriculture Organization of the United Nations to estimate calories in nonprocessed food purchases, assumes that processed calories are four times the price of nonprocessed calories, and uses the lower bound of the estimate of calories consumed in meals outside the home and given to nonhousehold members.

The results show a wide range of estimates of average calories consumed and the size of the food-insecure population (table 10). There is an average difference of approximately 280 calories purchased per capita between the High and Low estimates (column 3), which translates into a difference in the share of food-insecure households in the Indian sample of 16 percent (column 3). If we extrapolate to the entire country, the difference in the share of food-insecure households in India is 16.8 percent, which translates into 173 million people according to the 2001 Indian Census. The largest contributor to the difference between the High and Low estimates is the uncertainty about how many calories are contained in processed foods.

Table 11 presents the differences in the share of total households in India that are within various ranges of calorie purchases. The results indicate that the difference in average consumption between the High and Low calorie estimates in table 10 is accompanied by relatively large differences in households purchasing less than 1,500 and more than 2,400 calories per capita per day. Thus, although the primary sources of uncertainty in calorie consumption are largest for relatively richer households, households experiencing severe food insecurity are also affected.

Table 10. Comparison of high and low scenarios for calories consumed

	All sample households			Total country estimates		
	1	2	3	4	5	6
Variable	**High-calorie estimate**	**Low-calorie estimate**	**Difference**	**High-calorie estimate**	**Low-calorie estimate**	**Difference**
Average calorie consumption	2,471 (3.09)	2,191 (2.46)	280.2*** (1.52)	2,379 (4.88)	2,120 (4.20)	259.2*** (2.49)
Share of food-insecure households	.376 (.001)	.536 (.001)	-.160*** (.001)	.395 (.003)	.563 (.003)	-.168*** (.002)

Notes:

1. Estimates calculated using 122,888 households from the 61st round of the National Sample Survey. Population estimates are constructed using NSS multipliers to calculate population weights. Standard errors clustered at the district-level are presented in parentheses.
2. *** Denotes significance at the 1% level; ** denotes significance at the 5% level; * denotes significance at the 10% level.
3. Columns 1 and 4 calculate the estimate of calories purchased using the 'High Calorie' estimate described above and the share of households that purchased less than 2,100 calories. Columns 2 and 5 calculate the same totals using the 'Low Calorie' estimate. Columns 3 and 6 present the difference between the two, subtracting columns 2 and 5 from columns 1 and 4, respectively.

Source: Authors' calculations (see note 1).

Table 11. Changes to the depth of food security between the high and low scenarios for calories consumed

	Share of all Indian households in the food security range:		
	1	2	3
Food security range	High-calorie estimate	Low-calorie estimate	Difference (column 1 minus column 2)
Cal<1,500	.064 (.001)	.151 (.002)	-.087*** (.002)
1,500<Cal<1,800	.133 (.002)	.198 (.002)	-.065*** (.002)
1,800<Cal<2,100	.198 (.002)	.213 (.002)	-.015*** (.002)
2,100<Cal<2,400	.196 (.002)	.167 (.002)	.029*** (.002)
Cal>2,400	.408 (.003)	.270 (.002)	.138*** (.001)

Notes:

1. Estimates calculated using 122,888 households from the 61st round of the National Sample Survey. Estimates are constructed using NSS multipliers to calculate population weights. Standard errors clustered at the district level are presented in parentheses.
2. *** Denotes significance at the 1% level; ** denotes significance at the 5% level; * denotes significance at the 10% level.
3. The first column calculates the share of all Indian households in each range of total calories purchased using 'High Calorie' estimates described above. Column 2 calculates the share of households in each range of total calories purchased using the 'Low Calorie' estimates. Column 3 presents the difference between the two shares.

Source: Authors' calculations (see note 1).

Comparison with Direct Response Method Estimates of Household Food Security Status

A third methodology for assessing food security status is based on survey instruments that ask direct questions on food availability and food access for household members. This approach is used in the United States with a survey instrument, the Household Food Security Survey Module (HFSSM), which administers up to 18 questions to elicit information on food security conditions, with the data then used to place households on a food security scale (National Research Council, 2006). The U.S. Agency for International Development has developed a modification of this approach for assessing food security conditions in developing countries, using a simplified three-question module to place households on a 3-point Household Hunger Scale (HHS) of "Food Secure," "Food Insecure Without Hunger," and "Food Insecure With Hunger" categories (Deitchler et al., 2011).

The NSSO survey used for our study administers a single direct question on household food security status: "Do all members of your household get enough food every day?"[24] Responses to this question in the 2004/05 survey indicated that just 19 million people—or about 1.9 percent of the population—had inadequate access to food in at least 1 month of the previous year, sharply below what other information on India's food security would suggest. However, given the simplicity of the NSSO instrument, this 2004/05 result may not be a credible application of the direct response methodology.

Unfortunately, other Indian surveys using more complete versions of the HFSSM or HHS than the NSSO also do not provide results that can be readily compared with our national estimates based on aggregate production and consumption or household calorie consumption methods. Agarwal et al. (2009) surveyed 410 adult females in a Northeast Delhi slum using a 4-item variant of the HFSSM and found 51 percent of respondents to be food insecure. Gopichandran et al. (2010) studied 130 urban households in Vellore, Tamil Nadu using the HFSSM and found 61.5 percent of respondents to be food insecure with hunger and 13.1 percent to be food insecure without hunger. A survey of 282 low-income households in a backward area of Orissa reported by Nord et al. (2002), also using a variant of the HFSSM, found 92 percent of households to be food insecure and 57 percent to be food insecure with hunger. The narrow geographic scope, differing time periods, and nonrandom nature of the samples used in each of these studies make it impossible to compare the results using the direct response methodology with our results. A recent study provided validation of the internal consistency of results using the HHS methodology across several countries, but it did not provide external validation of HHS-based results against other methods and measures, such as household expenditure, food consumption, or nutritional status (Deitchler et al., 2010).

CONCLUSION

We find that estimates of the number of households classified as food insecure vary significantly with alternate plausible assumptions. Overall, when accounting for the three sources of error examined in this report, the share of India's population consuming less than 2,100 calories per day ranges from 39.5 percent (404 million people) to 56.3 percent (577 million people), for a difference of approximately 173 million people.[25] By contrast, USDA's Food Security Assessment for 2004/05 (Meade et al., 2006) estimated the world's total food-insecure population, using the same 2,100 calorie per day benchmark, at 775 million people and India's share at 217 million. Thus, the potential error in one country is equivalent to about 22 percent of the total food-insecure population of 70 developing countries in 2004/05. Although the low estimates derived here suggest the possibility that India's food situation could be better than indicated by some assessments based on household data, even the most conservative estimates in this analysis indicate that nearly 40 percent of the population is food insecure based on the 2,100 calorie per capita daily standard.

We are required to make a number of assumptions to estimate calories consumed per household member, even when using detailed household expenditure data. Assumptions regarding the amount of calories from each source, calories contained in nonprocessed and processed foods, and calories contained in meals eaten outside the household and given to nonhousehold members all have the potential to introduce error in estimates of calories purchased by households. In particular, the estimates of calories purchased are sensitive to assumptions accounting for calories contained in processed foods. These factors can result in estimates of calorie purchases that are different than the true values and that, in turn, affect the analysis of researchers concerned with the efficacy of programs designed to improve food security.

In estimating calories consumed in India—the country with the largest number of food-insecure individuals in the world—we find important characteristics in the consumption patterns among India's most vulnerable population. Although it is commonly thought that processed foods are important primarily for more affluent consumers in developing countries, the poorest households in India devote a nontrivial share of their purchases to such foods. Furthermore, relative to more affluent households, the poorest households purchase significantly cheaper sources of calories, primarily composed of grains, rather than a more diversified diet. This pattern is consistent with the persistently high rates of malnutrition in India.[26]

The approach in this study to estimating actual calorie consumption is in contrast to methods that assess food security by administering a survey module that asks respondents directly about the adequacy of food consumption for household members. The single question administered in the NSSO survey yielded a sharply lower estimate of the food-insecure population than other measures, but the instrument may be too simplistic to represent a credible application of this method. Other studies in India using the direct response approach all used localized, nonrandom samples in different years, so it is not possible to compare them with the national results using either the aggregate production and consumption or household consumption survey methods. Additional work may be needed to strengthen the direct response survey methodology used for national surveys in India, as well as to provide more external validation of results using the direct response approach against results using the other methods.

In addition to obfuscating the actual number of food-insecure people in the sample, the potential measurement error introduced in estimating calories consumed has important implications for any empirical analysis of calorie consumption. For example, some studies have found that calorie consumption in India has been decreasing over time (NSSO, 2007; Deaton and Dreze, 2008). Researchers have also noted, however, that consumption of processed foods and meals outside the household have been increasing (Kumar et al., 2007). Given that the data do not indicate the number of calories in those sources, it is also possible that, if calories from those sources have been systematically underestimated, calorie consumption might not actually be decreasing and could even be increasing for a subset of the population. Without a better understanding of the nutritional content of all sources of calories, it is very difficult even to identify trends in calorie consumption over time.[27]

These issues suggest that researchers and policymakers should rely on a number of different indicators of calorie consumption and food security beyond those based on household consumption surveys. Additional measures could include information on observable health indicators, such as emaciation or stunted growth, that are associated with low calorie consumption, as well as studies of aggregate food availability.[28] Finally, more attention might be given to strengthening survey enumeration methods to reduce key sources of error in computing calorie purchases.

Appendix 1. Conversion of Household Size to Effective Number of Young Adults

We used the following conversions, which were used by NSSO (2007), when converting each household member into the effective number of young adults:

Age in completed years	Male	Female
(1)	(2)	(3)
less than 1	0.43	0.43
1-3	.54	.54
4-6	.72	.72
7-9	.87	.87
10-12	1.03	.93
13-15	.97	.80
16-19	1.02	.75
20-39	1.00	1.71
40-49	.95	.68
50-59	.90	.64
60-69	.80	.51
70+	.70	.50

Appendix 2. Processed Food Items for Which Calorie Counts Could Not be Determined

The following is a list of food items (NSSO code in parentheses) for which only values of household expenditure are given or for which it is too difficult to estimate the number of calories purchased:

Other Rice Products (106), Other Wheat Products (114), Other Cereals (122), Other Pulse Products (153), Other Milk Products (167), Other Edible Oils (174), Others: birds, crab, oyster, tortoise, etc. (186), Other Vegetables (224), Other Fresh Fruits (247), Other Dry Fruits (257), Cold Beverages (295), Fruit Juice and Shake (296), Other Beverages (298), Biscuits (300), Salted Refreshment (301), Prepared Sweets (302), Cake/Pastry (304), Sauce (306), Other Processed Food (308).

References

Agarwal, Siddharth, Sethi, V., Gupta, P., Jha, M., Agnihotri, A. & Nord, M. (2009). "Experiential Household Food Insecurity in an Urban Underserved Slum of North India," *Food Security*, (1), 239–50.

Barrett, Christopher (2010). "Measuring Food Insecurity," *Science*, *327*(5967), 825-28.

Borjas, George (1980). "The Relationship between Wages and Weekly Hours of Work: The Role of Division Bias," *Journal of Human Resources*, *15*(3), 409-23.

Deaton, Angus & Subramanian, S. (1996). "The Demand for Food and Calories," *Journal of Political Economy*, *104*(1), 133-62.

Deaton, Angus & Dreze, J. (2008). "Food and Nutrition in India–Facts and Interpretations," *Economic and Political Weekly*, *44*(7), 42-65.

Deitchler, M., Ballard, T., Swindale, A. & Coates, J. (2010). "Validation of a Measure of Household Hunger for Cross-Cultural Use." Technical Report, FANTA-2, Food and Nutrition Technical Assistance Project, U.S. Agency for International Development.

Deitchler, M., Ballard, T., Swindale, A. & Coates, J. (2011). "Introducing a Simple Measure of Household Hunger for Cross-Cultural Use." Technical Note No. 12, FANTA-2, Food and Nutrition Technical Assistance Project, U.S. Agency for International Development.

Food and Agriculture Organization of the United Nations (FAO) (2011). *The State of Food Insecurity in the World 2011. Rome: FAO.*

Gopalan, C., Rama Sastri, B. V. & Balasubramanian, S. C. (1989). *Nutritive Value of Indian Foods.* Hyderabad: National Institute of Nutrition, Indian Council of Medical Research.

Gopichandran V., Claudius, P., Baby, L. S., Felinda, A. & Mohan, V. R. (2010). "Household food security in urban Tamil Nadu: a survey in Vellore," *National Medical Journal of India*, *23*(5), 278-80.

Hicks, Daniel (2009). "Consumption Volatility, Marketization, and Expenditure in Emerging Market Economies." Working Paper, Department of Economics, University of Oklahoma. Accessed January 2011, http://faculty-staff.ou.edu/H/Daniel.Hicks-1/.

Kumar, P., Mruthyunjaya, J. & Birthal, P. S. (2007). "Changing Consumption Patterns in South Asia," in P.K. Joshi, A. Gulati, and R. Cummings, Jr. (eds.), *Agricultural Diversification and Smallholders in South Asia.* New Delhi: Academic Foundation, 151-94.

Meade, Birgit, Rosen, S. & Shapouri, S. (2006). "Food Security Assessment, 2005." Outlook GFA-17, U.S. Dept. of Agriculture, Economic Research Service. Accessed January 2011, www.ers.usda. gov/publications/.

Meenakshi, J. V. & Vishwanathan, B. (2003). "Calorie Deprivation in Rural India," *Economic and Political Weekly*, *38*(4), 369-75.

Minhas, B. S (1991). "On Estimating the Inadequacy of Energy Intakes: Revealed Food Consumption Behaviour versus Nutritional Norms," *Journal of Development Studies*, *28*(1), 1-38.

National Research Council (2006). "Food Insecurity and Hunger in the United States: An Assessment of the Measure." Panel to Review U.S Department of Agriculture's Measurement of Food Insecurity and Hunger.

National Sample Survey Organization (2007). "Nutritional Intake in India 2004-2005." Ministry of Statistics and Program Implementation, Report No. 513. Accessed January 2011, http://mospi.nic.in/rept%20_%20pubn/513_final.pdf.

Nord, Mark, Satpathy, A. K., Raj, N., Webb, P. & Houser, R. (2002). "Comparing Household Survey-Based Measures of Food Insecurity across Countries: Case Studies in India, Uganda, and Bangladesh." Discussion Paper No. 7, Food Policy and Applied Nutrition Program, School of Nutrition Science and Policy, Tufts University.

Nord, Mark, Andrews, M. & Carlson, S. (2008). *Household Food Security in the United States, 2007*. ERR-66, U.S. Dept. of Agriculture, Economic Research Service. Accessed January 2011, www.ers.usda. gov/ publications/.

Saxena, N. C. (2011), "Hunger, Under-nutrition and Food Security in India." Working Paper 44, Chronic Poverty Research Center, Indian Institute of Public Administration. Accessed September 2011, http:// www. chronicpoverty.org/uploads/publication_files/CPRC-IIPA%2044. pdf.

Shapouri, Shahla, Rosen, S., Peters, M., Tandon, S., Gale, F., Mancino, L. & Bai, J. (2011). *International Food Security Assessment,* 2011-21. Outlook Report No. GFA-22, U.S. Dept. of Agriculture, Economic Research Service. Accessed September 2011, http://www.ers.usda.gov/Publications/gfa22/.

Tandon, S. & Landes, M. (2011). "The Sensitivity of Food Security in India to Alternate Estimation Strategies," *Economic and Political Weekly*, *46*(22), 92-99.

End Notes

[1] Surveys for 124,624 households were reported, but only 124,584 could be matched to probability weights, which are needed to estimate purchases for the entire Indian population from the sample. Furthermore, 48 households were excluded where household size could not be calculated. We also conducted robustness checks to exclude households with possible recording errors. In particular, the expenditure on processed food items listed in the appendix, which is a subset of total food expenditure, is greater than the total food expenditure for 1,041 households. These households have been excluded from the analysis. We also exclude sample households with implausibly high consumption and households where expenditures on nonprocessed food were equal to zero but consumption of nonprocessed calories was positive. In all, these adjustments exclude 1,648 of the original 124,536 households, leaving a total of 122,888 households.

[2] Probability weights are calculated using the multipliers provided by the NSSO, which are the number of households in the entire population represented by the household in the sample. We first multiply the multiplier by the size of the household, which gives us the number of people that are represented by the household members, and then divide by the total population to get the weights used in the calculations. Results are qualitatively identical if households are treated as the level of observation, with weights equal to the multiplier divided by the total number of Indian households estimated in the 2001 Indian census.

[3] See appendix 2 for a complete list of food items for which calorie information was too difficult to infer.

[4] Previous research suggests that meals consumed outside the household might be particularly important to poorer households and thus significantly affect estimates of calorie consumption (e.g., Minhas, 1991).

[5] See appendix 1 for a table describing how each household member translates into the effective number of young adults requiring 2,100 calories per day. In the sample of 122,888 households, the effective household size was 4.68. On the other hand, some studies report per capita consumption (e.g., Deaton and Subramanian, 1996; Deaton and Dreze, 2008), while others try to account for differing calorie requirements by counting each child as half a household member (e.g., Hicks, 2009). The household size in each of these scenarios is 4.94 and 4.88 respectively, which would result in a slightly smaller per capita consumption and higher food-insecure population when this consumption is compared with a consumption benchmark of 2,100 calories.

[6] The benchmark daily calorie consumption used in this analysis (2,100 calories per day) differs from NSSO (2007) estimates. NSSO uses different benchmarks for rural and urban households—2,400 calories/day for rural, 2,100 calories/day for urban. Previous research has suggested that the actual benchmark chosen significantly changes estimates of food insecurity (e.g., Meenakshi and Vishwanathan, 2003). However, the primary point of this analysis is to demonstrate the changes in estimates of food security in response to changing assumptions. When different benchmarks are used for rural and urban areas, the main result stands that there is a large difference in the estimated number of food-insecure individuals.

[7] A rural household is classified as affluent by the NSSO if it owns a motor vehicle, mechanized farm equipment, or electrical appliances, if any family member is a highly paid professional such as a doctor, if the household owns a sufficient amount of land, or if it owns at least 10 head of cattle and/or buffalo.

[8] It is interesting to note that, although richer households consumed more meat, the number of meat calories is much smaller than any other food category and has a very small standard error. Thus, richer households generally substitute more towards nonmeat food groups.

[9] These estimates of sources of nonprocessed foods are similar to estimates presented in the NSSO (2007).

[10] See appendix 2 for a complete list of food items that are included as "Processed Foods."

[11] Control variables include the amount of money spent on paan (a nut and spice mixture wrapped in betel leaf, for chewing), tobacco, intoxicants; fuel; the monthly per capita expenditure for the household; the effective number of young adults in the household; an indicator for whether the household is rural, and an indicator for whether the household is reporting insufficient food sources during the year.

[12] This approach differs from the methodology reported by NSSO (2007), which provides the number of calories used to estimate the calories contained in meals eaten outside of the household, meals given to nonhousehold members, and calories contained in processed foods. However, NSSO does not provide an explanation of the source of these values.

[13] The bounds of the confidence interval were calculated by multiplying the estimated standard error of the coefficient estimate (7.84) by 1.96 and adding and subtracting the value from the coefficient estimate.

[14] The bounds of the confidence interval are again calculated by multiplying the estimated standard error of the coefficient estimate (225.6) by 1.96 and adding and subtracting the value from the coefficient estimate.

[15] These estimations suggest considerable uncertainty in the calorie content of whole meals. This is especially true of meals received outside the home. However, this pattern suggests that the confidence interval should be even larger than suggested above and bolsters the argument that uncertainty in calorie content of these types of meals leads to uncertainty in calories available for consumption by households.

[16] There is some uncertainty as to what level of calorie purchases will actually result in a consumption of 2,100 calories. Wastage in calories purchased would imply that purchasing 2,100 calories would result in a lower level of consumption. However, the point of this analysis is to show how the assessment changes when assumptions change rather than simply reporting the raw numbers of the food insecure. Furthermore, the choice of 2,100 calories purchased can therefore be interpreted as more severe food insecurity than 2,100 calories consumed.

[17] This estimate is derived by multiplying the estimate of the share of all Indian households purchasing less than 2,100 calories by the number of households in the 2001 Indian Census, multiplied by the average household size in the 2001 Indian Census.

[18] There are no significant differences in the amount of measurement error in rural and urban households, and thus for simplicity, we report estimates for the sample pooled across rural and urban sectors.

[19] This concordance is available online at http://faostat.fao.org/site/368/DesktopDefault.aspx?PageID=368; we accessed the site in August 2010. For food items in the NSSO survey that have no link to the food items in the FAO list, the calorie value from NVIF was used to make the two sources of calorie information as similar as possible.

[20] To illustrate the difference between the two sources: The NSSO survey records purchases of 9 different types of pulses, NVIF reports nutritional information for 19 distinct types of pulses, and FAO reports nutritional information for only 3 types. Thus, 9 of the 19 pulse items reported in NVIF are matched to the NSSO categories, while 1 of the 3 FAO calorie values are matched to each of the 9 different food categories. In both instances, the calorie counts are relatively close, with the average calories in a kilogram of pulses from the NVIF and FAO measured at 3,418 and 3,511.4, respectively.

[21] Control variables include the amount of money spent on paan, tobacco, intoxicants, and fuel, the monthly per capita expenditure for the household, the effective number of young adults in the household, an indicator for whether the household is rural, and an indicator for whether the household is reporting insufficient food sources during the year.

[22] The primary driver of this difference appears to be the calories contained in certain types of cereals and pulses. In each food category, the NVIF has approximately 300 more calories contained in a particular type of cereal and approximately 200 more calories in a particular type of pulse. These two food items account for a large amount of calories purchased, and thus NVIF results in a larger estimate of calorie consumption.

[23] Euromonitor International provides a list of processed foods along with prices and quantities available in India. The data were accessed at http://portal.euromonitor.com/Portal/ResultsList.aspx on 10/12/2010. Analyzing just a few of many options available to consumers in the types of processed foods listed in appendix 2, we found that calories purchased per rupee averaged 47.9, with a minimum of 4.7 and a maximum of 348.8. The hypotheses that processed foods are the same price as nonprocessed foods (269 calories per rupee, on average) and that processed foods are four times as expensive (67.25 calories per rupee, on average) are within this range.

[24] Response choices are: (1) Yes: Every month of the year; (2) Some months of the year; (3) No: no month of the year. Households choosing either (2) or (3) are counted as food insecure.

[25] The population estimates are based on sample weights linked to India's 2001 Indian Census and are not strictly comparable to the USDA estimates for 2004/05.

[26] Malnutrition rates estimated by the World Health Organization are available at http://www.who.int/nutgrowth-db/database/countries/en/ (Accessed January 2011).

[27] There are other concerns in using calories purchased per household member as a dependent variable in any empirical analysis. In particular, in the construction of the variable, we used household size, the amount of money used to purchase food, and a number of other household-level variables. We cannot consistently estimate the relationship between total calories purchased as the dependent variable and any of these variables, or even variables constructed using them, given the possibility of division bias (Borjas, 1980). In these instances, even if the true correlation between the variables is precisely zero, the mechanics of constructing the variables can cause a spurious correlation.

[28] In addition, policymakers should focus on the response of food security and calories consumed to policy interventions and income and production shocks that cause exogenous variation in the independent variables of interest.

INDEX

V

W

Y